T0236196

Operator Theory: Advances and Applications

Vol. 209

Founded in 1979 by Israel Gohberg

Stability of Operators and Operator Semigroups

Tanja Eisner

Birkhäuser

Author:

Tanja Eisner
Arbeitsbereich Funktionalanalysis
Mathematisches Institut
Auf der Morgenstelle 10
72076 Tübingen
Germany
e-mail: talo@fa.uni-tuebingen.de

2010 Mathematics Subject Classification: 47A10, 47D06, 47A35, 28D05

Bibliographic information published by Die Deutsche Bibliothek.
Die Deutsche Bibliothek lists this publication in the Deutsche Nationalbibliografie;
detailed bibliographic data is available in the Internet at http://dnb.ddb.de

© 2010 Springer Basel AG
Softcover reprint of the hardcover 1st edition 2010
P.O. Box 133, CH-4010 Basel, Switzerland
Part of Springer Science+Business Media
Printed on acid-free paper produced from chlorine-free pulp. TCF∞

ISBN 978-3-0348-0311-3   ISBN 978-3-0346-0195-5 (eBook)
9 8 7 6 5 4 3 2 1   www.birkhauser-science.ch

Contents

Introduction

The real understanding involves, I believe, a synthesis of the discrete and continuous ...

L. Lovász, Discrete and Continuous: two sides of the same?

Evolving systems can be modelled using a discrete or a continuous time scale. The discrete model leads to a map φ and its powers φ^n on some state space Ω, while the continuous model is given by a (semi)flow $(\varphi_t)_{t \geq 0}$ on Ω satisfying $\varphi_0 = Id$ and $\varphi_{t+s} = \varphi_t \varphi_s$. In many situations, the state space Ω is a Banach space and the maps are linear and bounded. In this case, we will use the notations T and $T(t)$ instead of φ and φ_t.

In this book, our focus is on the various concepts of stability of such linear systems, where by stability we mean convergence to zero of $\{T^n\}_{n=1}^{\infty}$ or $(T(t))_{t \geq 0}$ in a sense to be specified. This property is crucial for study of the qualitative behaviour of dynamical systems, even in the non-linear case. We adopt the general philosophy of a parallel treatment of both the discrete and the continuous case, systematically comparing methods and results. We try to give a reasonably complete picture mentioning (most of) the relevant results, a strategy that has, by the way, helped us to identify a number of natural open problems. However, we do not discuss the case of positive operators and semigroups on Banach lattices, referring the reader to, e.g., Nagel (ed.) [196] or the recent monograph of Emel'yanov [75] for this topic. We also tried to minimise overlap with the monographs of van Neerven [204] on asymptotics in the continuous case and Müller [191] in the discrete case. Instead we emphasise the connections of stability in operator theory to its analogues in ergodic theory and harmonic analysis.

In the following we summarise the content of the book.

Chapter I gives an overview on some functional analytic tools needed later. Besides the classical decomposition theorems of Jacobs–Glicksberg–de Leeuw for compact semigroups, we recall the powerful notion of the cogenerator of a C_0-semigroup. Finally, we present one of our main concepts for the investigation of stability of C_0-semigroups, the Laplace inversion formula, which can also be seen as an extension of the Dunford functional calculus for exponential functions.

In Chapter II we investigate the behaviour of the powers of a bounded linear operator on a Banach space. As a first step, we study power boundedness. Here and later, behaviour of the resolvent of the operator near the unit circle plays a crucial role. In particular, we give a resolvent characterisation for power boundedness on Hilbert spaces using the L^2-norm of the resolvent on circles with radius greater than 1. We further study the related notion of polynomial boundedness which is, surprisingly, easier to characterise than power boundedness.

Stability is the topic in the rest of the chapter. While uniform stability is characterised by the spectral radius, we discuss strong stability and give various characterisations of it, both classical and recent. When turning to weak stability, we encounter a phenomenon, well-known in ergodic theory and described by Katok and Hasselblatt [143, p. 748] as follows[1]

> "... It [mixing] is, however, one of those notions, that is easy and natural
> to define but very difficult to study...".

Thus very few results concerning weak stability are known, a sufficient condition in terms of the resolvent being one of the exceptions.

We then introduce the concept of almost weak stability and give various equivalent conditions. From this we see that almost weak stability, while looking artificial, is much easier to characterise than weak stability. In particular, if the operator has relatively compact orbits, then almost weak stability is equivalent to "no point spectrum on the unit circle".

Almost weak stability turns out to be fundamentally different from weak stability, as we see in Section 5. We show that a "typical" (in the sense of Baire) unitary operator, a "typical" isometry, and a "typical" contraction on a separable Hilbert space are almost weakly, but not weakly, stable. This is an operator theoretic counterpart to the classical theorems of Halmos [116] and Rohlin [221] on weakly and strongly mixing transformations on a probability space.

We close the chapter by characterising stability and power boundedness of operators on Hilbert spaces via Lyapunov equations.

In Chapter III we turn our attention to the time continuous case and consider C_0-semigroups $(T(t))_{t\geq 0}$ on Banach spaces. As in the previous chapter we discuss boundedness and stability of $(T(t))_{t\geq 0}$ and try to characterise these properties by the generator, its spectrum and resolvent. We start with boundedness and discuss in particular the quite recent characterisation of a bounded C_0-semigroup on a Hilbert space by Gomilko [104] and Shi, Feng [231]. Their integrability condition for the resolvent on vertical lines (Theorem 1.11) is the key to the resolvent approach to boundedness and stability, as first introduced by van Casteren [45, 46]. We then characterise generators of polynomially bounded semigroups (i.e., semigroups growing not faster than a polynomial) in terms of various resolvent conditions.

---

[1] Mixing corresponds to weak stability.

A discussion of uniform exponential stability of C_0-semigroups follows, a notion which is more difficult to characterise than its discrete analogue. Besides Gearhart's generalisation to Hilbert spaces of Liapunov's stability theorem and the Datko–Pazy theorem, we extend Gearhart's theorem to Banach spaces.

Strong stability is the subject of the following section containing the classical results of Sz.-Nagy and Foiaş, Lax and Phillips and the more recent theorem by Arendt, Batty [9], Lyubich and Vũ [180]. We then discuss the resolvent approach developed by Tomilov [243].

Weak stability and its characterisations, partly classical and partly quite recent, is the next topic. We emphasise that, as in the discrete case, weak stability is much less understood than its strong and uniform analogues, still leaving many open questions. For example, it is not clear how to characterise weak stability in terms of the resolvent of the generator.

Next, we look at almost weak stability (Definition 5.3), which is closely related to weak stability but occurs much more frequently. We give various equivalent conditions, analogous to the discrete case, and present a concrete example of an almost weakly, but not weakly, stable semigroup. We finally present category theorems analogous to the discrete ones stating that a "typical" (in the sense of Baire) unitary C_0-group as well as a "typical" isometric C_0-semigroup on a separable infinite-dimensional Hilbert space are almost weakly, but not weakly, stable.

Characterisation of stability and boundedness of C_0-semigroups on Hilbert spaces via Lyapunov equations is the subject of the last section.

Chapter IV relates our stability concepts to weakly and strongly mixing transformations and flows in ergodic theory and (via the spectral theorem) to Rajchman and non-Rajchman measures in harmonic analysis. In addition, "typical" behaviour of contractive operators and semigroups on Hilbert spaces is studied using the notion of rigidity, extending the corresponding results in the previous two chapters.

In Chapter V we build bridges between discrete systems $\{T^n\}_{n=0}^{\infty}$ and continuous systems $(T(t))_{t\geq 0}$. We start by embedding a discrete system into a continuous one. More precisely, we ask for which operators T there exists a C_0-semigroup $(T(t))_{t\geq 0}$ with $T = T(1)$. We discuss some spectral conditions for embeddability and give classes of examples for which such a C_0-semigroup does or does not exist. The general embedding question is still open as well as its analogues in ergodic and in measure theory. In Section 2 we discuss the connection between a C_0-semigroup $T(\cdot)$ and its cogenerator V and its induced discrete system $\{V^n\}_{n=0}^{\infty}$. On Hilbert spaces there are great similarities between asymptotic properties of the two systems. We give classical and more recent results, discuss difficulties, examples, and open questions.

Acknowledgment. My deepest thanks go to Rainer Nagel, whose strong encouragement and support, deep mathematical insight and permanent optimism has been an invaluable help.

I am very grateful to Wolfgang Arendt, Ralph Chill, Bálint Farkas, Jerome Goldstein, Sen-Zhong Huang, Dávid Kunszenti-Kovács, Yuri Latushkin, Mariusz Lemańczyk, Jan van Neerven, Werner Ricker, Yuri Tomilov, Ulf Schlotterbeck, András Serény, Jaroslav Zémanek and Hans Zwart for helpful and interesting discussions and comments on this book.

The cordial working atmosphere in the AGFA group (Arbeitsgemeinschaft Funktionalanalysis) in Tübingen has been very stimulating and helpful. Special thanks go to András Batkai, Fatih Bayazit, Ulrich Groh, Retha Heymann, Bernd Klöss, Dávid Kunszenti-Kovács, Felix Pogorzelski, Ulf Schlotterbeck, Marco Schreiber, and Pavel Zorin for reading preliminary versions of the manuscript.

Support by the Dr. Meyer-Struckmann Stiftung (via the Studienstiftung des deutschen Volkes), the European Social Fund and the Ministry Of Science, Research and the Arts Baden-Württemberg is gratefully acknowledged.

I deeply thank my parents for their hearty encouragement and permanent support, as well as Anna M. Vishnyakova for teaching me how to become a mathematician.

Chapter I

Functional analytic tools

In this chapter we introduce some functional analytic tools needed later.

We start with the theory of general compact semigroups and the decomposition theorem of Jacobs–Glicksberg–de Leeuw for (discrete or continuous) operator semigroups. If applicable it states that an operator semigroup can be decomposed into a "reversible" and a "stable" part. Since the reversible part corresponds to an action of a compact group and can be described using harmonic analysis, it remains to study the stable part, the main goal of this book.

We then discuss mean ergodicity of operators and semigroups and some concepts from semigroup theory such as the cogenerator and the inverse Laplace transform for C_0-semigroups.

1 Compact semigroups

1.1 Preliminaries

We first recall some facts concerning the weak topology and weak compactness in Banach spaces and refer to Dunford, Schwartz [63, Sections V.4-6], Rudin [225, Chapter 3] and Schaefer [226, Section IV.11] for details.

We begin with the Eberlein–Šmulian theorem characterising weak compactness in Banach spaces.

Theorem 1.1 (Eberlein–Šmulian). *For subsets of a Banach space, weak compactness and weak sequential compactness coincide.*

Next, we recall the following consequence of the Banach–Alaoglu theorem.

Theorem 1.2. *A Banach space is reflexive if and only if its closed unit ball is weakly compact.*

In particular, every bounded set of a Banach space is relatively weakly compact if and only if the space is reflexive.

Another important property of weakly compact sets is expressed by the Kreĭn–Šmulian theorem.

Theorem 1.3 (Kreĭn–Šmulian). *The closed convex hull of a weakly compact subset of a Banach space is weakly compact.*

The following theorem characterises metrisability of the weak topology.

Theorem 1.4. *The weak topology on the closed unit ball of a Banach space X is metrisable if and only if the dual space X' is separable.*

We recall that separability of X' implies separability of X, while the converse does not always hold. In addition, separable Banach spaces have a nice metrisability property with respect to the weak topology.

Theorem 1.5. *On a separable Banach space, the weak topology is metrisable on weakly compact subsets.*

We now introduce the basic notations to be used in the following.

For a linear operator T we denote by $\sigma(T)$, $P_\sigma(T)$, $A_\sigma(T)$, $R_\sigma(T)$, $r(T)$, $\rho(T)$ and $R(\lambda, T)$ its spectrum, point spectrum, approximative point spectrum and residual spectrum, spectral radius, resolvent set and resolvent at the point $\lambda \in \rho(T)$, respectively. Here, the approximative point spectrum and residual spectrum are defined as

$$A_\sigma(T) := P_\sigma(T) \cup \{\lambda \in \mathbb{C} : \text{rg}(\lambda I - T) \text{ is not closed in } X\}$$

$$= \left\{\lambda \in \mathbb{C} : \exists \{x_n\} \subset X \text{ such that } \|x_n\| = 1 \text{ and } \lim_{n \to \infty} \|Tx_n - \lambda x_n\| = 0\right\},$$

$$R_\sigma(T) := \{\lambda \in \mathbb{C} : \text{rg}(\lambda I - T) \text{ is not dense in } X\}.$$

For more on these concepts see e.g. Conway [52, §VII.6] and Engel, Nagel [78, pp. 241–243].

For a C_0-semigroup $(T(t))_{t \geq 0}$ we often write $T(\cdot)$, and denote its growth bound by $\omega_0(T)$. The spectral bound $s(A)$ of an (in general unbounded) operator A is

$$s(A) := \sup\{\text{Re}\,\lambda : \lambda \in \sigma(A)\}$$

and its *pseudo-spectral bound* (also called *abscissa of the uniform boundedness of the resolvent*) is

$$s_0(A) := \inf\{a \in \mathbb{R} : R(\lambda, A) \text{ is bounded on } \{\lambda : \text{Re}\,\lambda > a\}\}.$$

Recall that for a C_0-semigroup $T(\cdot)$ with generator A the relations

$$s(A) \leq s_0(A) \leq \omega_0(T)$$

hold, while both inequalities can be strict, see e.g. Engel, Nagel [78, Example IV.3.4] and van Neerven [204, Example 4.2.9], respectively. For other growth bounds and their properties see van Neerven [204, Sections 1.2, 4.1, 4.2].

1.2   Relatively compact sets in $\mathcal{L}_\sigma(X)$

In this subsection we characterise relatively weakly compact sets of operators and give some important examples. To this purpose we denote by $\mathcal{L}_\sigma(X)$ the space of all bounded linear operators on a Banach space X endowed with the weak operator topology.

The following characterisation goes back to Grothendieck.

Lemma 1.6. *For a set of operators $\mathcal{T} \subset \mathcal{L}(X)$, X a Banach space, the following assertions are equivalent.*

(a) *$\mathcal{T}$ is relatively compact in $\mathcal{L}_\sigma(X)$.*

(b) *$\mathcal{T}x := \{Tx : T \in \mathcal{T}\}$ is relatively weakly compact in X for all $x \in X$.*

(c) *$\mathcal{T}$ is bounded, and $\mathcal{T}x$ is relatively weakly compact in X for all x in some dense subset of X.*

Proof. We follow Engel, Nagel [78, pp. 512–514].

The implication (a)$\Rightarrow$(b) follows directly from the continuity of the mapping $T \mapsto Tx$ for every $x \in X$. The converse implication follows from the inclusion

$$(\mathcal{T}, \mathcal{L}_\sigma(X)) \subset \prod_{x \in X} (\overline{\mathcal{T}x}^\sigma, \sigma)$$

and Tychonoff's theorem.

The uniform boundedness principle yields the implication (b)$\Rightarrow$(c).

(c)$\Rightarrow$(b): Take $x \in X$ and $\{x_n\}_{n=1}^\infty \subset D$ converging to x, where D denotes the dense subset of X from (c). By the Eberlein–Šmulian theorem (Theorem 1.1), it is enough to show that every sequence $\{T_n x\}_{n=1}^\infty \subset \mathcal{T}x$ has a weakly convergent subsequence.

Take a sequence $\{T_n x\}_{n=1}^\infty \subset \mathcal{T}x$. For $x_1 \in D$, there exists a subsequence $\{n_{k,1}\}_{k=1}^\infty$ and a vector $z_1 \in X$ such that weak-$\lim_{k \to \infty} T_{n_{k,1}} x_1 = z_1$. Analogously, for $x_2 \in D$ there exists a subsequence $\{n_{k,2}\}_{k=1}^\infty$ of $\{n_{k,1}\}_{k=1}^\infty$ and a vector $z_2 \in X$ such that weak-$\lim_{k \to \infty} T_{n_{k,2}} x_2 = z_2$, and so on. By the standard diagonal procedure there exists a subsequence which we denote by $\{n_k\}_{k=1}^\infty$ such that weak-$\lim_{k \to \infty} T_{n_k} x_m = z_m$ for every $m \in \mathbb{N}$. We have

$$\|z_n - z_m\| \le \sup_{k \in \mathbb{N}} \|T_{n_k}(x_n - x_m)\| \le C\|x_n - x_m\|$$

for $C = \sup\{\|T\| : T \in \mathcal{T}\}$ and all $n, m \in \mathbb{N}$. So $\{z_n\}_{n=1}^\infty$ is a Cauchy sequence and therefore converges to some $z \in X$. By the usual 3ε–argument we obtain weak-$\lim_{k \to \infty} T_{n_k} x = z$. $\square$

We now give some examples of relatively weakly compact subsets of operators.

Example 1.7.

(a) On a reflexive Banach space X, any norm bounded family $\mathcal{T} \subseteq \mathcal{L}(X)$ is relatively weakly compact by the Banach–Alaoglu theorem.

(b) Let $\mathcal{T} \subseteq \mathcal{L}(L^1(\mu))$ be a norm bounded subset of positive operators on the Banach lattice $L^1(\mu)$, and suppose that $Tu \leq u$ for some μ-almost everywhere positive $u \in L^1(\mu)$ and every $T \in \mathcal{T}$. Then $\mathcal{T}$ is relatively weakly compact since the order interval $[-u, u]$ is weakly compact, $\mathcal{T}$-invariant and generates a dense subset (see Schaefer [227, Thm. II.5.10 (f) and Prop. II.8.3]).

(c) Let S be a semitopological semigroup, i.e., a (multiplicative) semigroup S which is a topological space such that the multiplication is separately continuous (see Definition 1.8 below). Consider the space $C(S)$ of bounded, continuous (real- or complex-valued) functions over S. For $s \in S$ define the corresponding rotation operator $(L_s f)(t) := f(s \cdot t)$. A function $f \in C(S)$ is said to be *weakly almost periodic* if the set $\{L_s f : s \in S\}$ is relatively weakly compact in $C(S)$, see Berglund, Junghenn, Milnes [33, Def. 4.2.1]. The set of weakly almost periodic functions is denoted by $WAP(S)$. If S is a *compact* semitopological semigroup, then $C(S) = WAP(S)$ holds, see [33, Cor. 4.2.9]. This means that for a compact semitopological semigroup S the set $\{L_s : s \in S\}$ is always relatively weakly compact in $\mathcal{L}(C(S))$. We come back to this example in the proof of Theorem II.4.1, in Example II.4.10 as well as in the proof of Theorem III.5.1 and in Example III.5.9.

1.3 Compact semitopological semigroups

In this subsection we present a very general and flexible approach to the qualitative theory of operators and C_0-semigroups. We will use the theory of compact semigroups and refer to Engel, Nagel [78, Section V.2] and Krengel [154, Section 2.2.4].

We call a set $\mathcal{S}$ with an associative multiplication $\cdot$ an *abstract semigroup* $(\mathcal{S}, \cdot)$. If the semigroup operation is fixed, we write only $\mathcal{S}$ and st instead of $(\mathcal{S}, \cdot)$ and $s \cdot t$.

Definition 1.8. An abstract semigroup $\mathcal{S}$ is called a *semitopological semigroup* if $\mathcal{S}$ is a topological space such that the multiplication is separately continuous, i.e., such that the maps $s \mapsto st$ and $s \mapsto ts$ are continuous for every $t \in \mathcal{S}$. If in addition $\mathcal{S}$ is compact, we call $\mathcal{S}$ a *compact semigroup*. A subset $J \subset \mathcal{S}$ is called an *ideal* if $J\mathcal{S} \subset J$ and $\mathcal{S}J \subset J$ hold.

The following easy lemma shows that separate continuity of the multiplication is enough to preserve the semigroup property and commutativity under the closure operation.

Lemma 1.9. *Let $\mathcal{S}$ be a semitopological semigroup and $\mathcal{T} \subset \mathcal{S}$. If $\mathcal{T}$ is a subsemigroup, then so is $\overline{\mathcal{T}}$. Moreover, if $\mathcal{T}$ is commutative, then so is $\overline{\mathcal{T}}$.*

We now state the classical theorem describing the structure of commutative compact semigroups.

Theorem 1.10 (Structure of compact semigroups). *Let S be a compact commutative semitopological semigroup. Then there exists a unique minimal ideal K in S which is a compact topological group (i.e., the multiplication and the inverse maps are continuous from $K \times K \to K$ and $K \to K$, respectively) and satisfies $K = qS$ for the unit element q of K.*

Proof. Observe first that, for closed ideals $J_1, \ldots, J_n$ in S, $\cap_{j=1}^n J_j$ is a nonempty closed ideal containing $J_1 J_2 \cdots J_n$. By compactness of S we conclude that

$$K := \bigcap_{J \text{ closed ideal}} J$$

is a nonempty closed ideal in S.

We now show that K is the minimal ideal. Let J be an arbitrary ideal in S and $s \in J$. We observe that $J_s := sS$ is an ideal as well and is closed as the image of the compact set S under the multiplication by s. Hence, $K \subset J_s \subset J$ and K is a minimal ideal. Uniqueness follows by the above representation of K.

We now check that K is a group. Take $s \in K$. Then $sK \subset K$ is an ideal as well, hence

$$sK = K \tag{I.1}$$

by the minimality of K. Therefore there exists $q \in K$ with $sq = s$. We show that q is the unit in K. Indeed, for an arbitrary $r \in K$ there exists by (I.1) some $r' \in K$ with $sr' = r$. So by commutativity of S we have $rq = sr'q = r's = r$ which implies that q is the unit element in K. Existence of the inverse follows again from (I.1).

The equality $K = qS$ follows from minimality of K and the ideal property of qS.

We now invoke the Ellis theorem (see, e.g., Namioka [201] or Glasner [96, pp. 33–37]) stating that each compact semitopological group is in fact a topological group. $\qquad\square$

1.4 Abstract Jacobs–Glicksberg–de Leeuw decomposition for operator semigroups

We now consider the case where S is a subsemigroup of $\mathcal{L}(X)$ (X a Banach space) considered with the usual multiplication. The typical examples for topologies making S a semitopological semigroup are the weak, strong and norm operator topologies.

If S is commutative and compact with respect to one of these topologies, the idempotent element q from Theorem 1.10 is a projection, hence

$$X = \ker q \oplus \operatorname{rg} q$$

and both subspaces are $\mathcal{S}$-invariant. Since the unit q of the minimal ideal $\mathcal{K} = q\mathcal{S}$ is zero on $\ker q$ and the identity operator on $\mathrm{rg}\, q$, we have

$$\mathcal{K} = \{0\} \oplus \mathcal{S}|_{\mathrm{rg}\, q}.$$

Therefore, the semigroup $\mathcal{S}$ satisfies

$$0 \in \mathcal{S}|_{\ker q} \text{ and } \mathcal{S}|_{\mathrm{rg}\, q} \text{ is a compact topological group.}$$

These properties describe the structure of commutative compact semigroups of operators on a Banach space. We state it for the weak operator topology only. Here and below, Γ denotes the unit circle in $\mathbb{C}$.

Theorem 1.11 (Abstract Jacobs–Glicksberg–de Leeuw decomposition for operator semigroups). *Let X be a Banach space and let $\mathcal{T} \subset \mathcal{L}(X)$ be a commutative, relatively weakly compact semigroup. Then $X = X_r \oplus X_s$, where*

$$X_r := \overline{\mathrm{lin}}\{x \in X : \text{ for every } T \in \mathcal{T} \text{ there exists } \gamma \in \Gamma \text{ such that } Tx = \gamma x\}$$

$$X_s := \{x \in X : 0 \text{ is a weak accumulation point of } \mathcal{T}x\}.$$

Proof. Denote by $\mathcal{S}$ the closure of $\mathcal{T}$ in the weak operator topology. Then $\mathcal{S}$ is a compact commutative semitopological semigroup, see Lemma 1.9. Denote by $\mathcal{K}$ its minimal ideal and by q the unit in $\mathcal{K}$. We show that $X_s = \ker q$ and $X_r = \mathrm{rg}\, q$.

Part 1: $X_s = \ker q$.

We observe first that, by continuity of the mapping $T \mapsto Tx$, $\mathcal{L}_\sigma(X) \to (X, \sigma)$, we have $\overline{\mathcal{T}x}^\sigma = \mathcal{S}x$ for every $x \in X$.

To show $\ker q \subset X_s$ take x with $qx = 0$. Then $0 \in \mathcal{S}x = \overline{\mathcal{T}x}^\sigma$ and hence $x \in X_s$. For the converse inclusion take $x \in X_s$. Then $0 \in \overline{\mathcal{T}x}^\sigma = \mathcal{S}x$ and therefore there exists an operator $R \in \mathcal{S}$ with $Rx = 0$ and hence $qRx = 0$ as well. Since $qR \in \mathcal{K}$ and $\mathcal{K}$ is a group, there exists an operator $R' \in \mathcal{K}$ with $R'qR = q$. This implies $qx = R'qRx = 0$.

Part 2: $X_r = \mathrm{rg}\, q$.

We first prove "$\subset$". Take $x \in X$ with $\mathcal{T}x \subset \Gamma \cdot x$. Then $\mathcal{S}x \subset \Gamma \cdot x$ as well and hence $qx = \gamma x$ for some $\gamma \in \Gamma$. (Indeed $\gamma = 1$ by $q^2 = q$.) Therefore, $x \in \mathrm{rg}\, q$. By density of such vectors in X_r and closedness of $\mathrm{rg}\, q$ we have $X_r \subset \mathrm{rg}\, q$.

The proof of $\mathrm{rg}\, q \subset X_r$ uses some basic facts on compact abelian groups, see e.g. Hewitt, Ross [127].

Denote by m the Haar measure on $\mathcal{K}$, i.e., the unique invariant probability measure with respect to multiplication. (For an elegant functional-analytic proof of its existence for compact groups see Izzo [133].) For a continuous character, i.e., a multiplicative continuous function $\chi : \mathcal{K} \to \Gamma$, define

$$P_\chi x := \int_\mathcal{K} \overline{\chi(S)} Sx \, dm(S)$$

weakly. (Note that the function $S \mapsto \overline{\chi(S)}\langle Sx, y\rangle$ is continuous.) We obtain an operator $P_\chi : X \to X''$ and show that actually $P_\chi : X \to X$. Indeed, for a fixed $x \in X$ the set $M := \{\overline{\chi(S)}Sx, S \in \mathcal{K}\}$ is weakly compact in X. Since the closed convex hull of a weakly compact set is again weakly compact by the Kreĭn–Šmulian theorem (see Theorem 1.3), $P_\chi x$ belongs to the image of $\overline{\text{co}}\{M\}$ in X'' under the canonical embedding and hence represents an element of X. So $P_\chi \in \mathcal{L}(X)$ with $\|P_\chi\| \leq \sup_{S \in \mathcal{K}} \|S\|$.

We now show that $\text{rg}(P_\chi) \subset \{x : \mathcal{T}x \subset \Gamma x\}$. Take $R \in \mathcal{K}$. By the invariance of m and multiplicativity of χ,

$$RP_\chi x = \int_K \overline{\chi(S)}RSx\, dm(S) = \chi(R) \int_K \overline{\chi(RS)}RSx\, dm(S),$$

$$= \chi(R) \int_K \overline{\chi(S)}Sx\, dm(S) = \chi(R)P_\chi x$$

holds for every $x \in X$, i.e., $RP_\chi = \chi(R)P_\chi$. For $R := q$ this means $qP_\chi = \chi(q)P_\chi = P_\chi$, hence $\text{rg}(P_\chi) \subset \text{rg}\, q$. Take now $S \in \mathcal{T}$ and define $R := Sq$. Then we have $SP_\chi = SqP_\chi = \chi(Sq)P_\chi$ and therefore

$$\text{rg}(P_\chi) \subset \{x : Sx = \chi(Sq)x \text{ for every } S \in \mathcal{T}\} \tag{I.2}$$

and in particular $\text{rg}(P_\chi) \subset X_r$.

It remains to show that

$$X_r \subset \overline{\text{lin}} \bigcup_\chi \text{rg}\, P_\chi.$$

To this purpose, take $y \in X'$ vanishing on $\text{rg}\, P_\chi$ for every character χ and show that y vanishes on $\text{rg}\, q$. By assumption $\int_K \overline{\chi(S)}\langle Sx, y\rangle dm(S) = 0$ for every character χ and $x \in X$. By the continuity of the integrand and by the fact that the linear hull of all characters is dense in $L^2(K, m)$ (see e.g. Hewitt, Ross [127, Theorem 22.17]), we obtain $\langle Sx, y\rangle = 0$ for every $S \in \mathcal{K}$. In particular, $\langle qx, y\rangle = 0$ for every x, so y vanishes on $\text{rg}\, q$ and the theorem is proved. $\qquad\square$

The subspaces X_r and X_s in the above decomposition are called *reversible* and *stable* subspaces, respectively.

Remark 1.12. A function $\lambda : \mathcal{T} \to \mathbb{C}$ is called an *eigenvalue function* if there exists $0 \neq x \in X$ such that $Tx = \lambda(T)x$ holds for every $T \in \mathcal{T}$. By (I.2) above, we see that every character χ with $P_\chi \neq 0$ defines an eigenvalue function $\lambda : \mathcal{K} \to \Gamma$ by $\lambda(T) = \chi(Tq)$ with eigenvectors in $\text{rg}\, P_\chi$. Conversely, if for some $x \neq 0$ the equalities $Tx = \lambda(T)x$ with $\lambda(T) \in \Gamma$ hold for every $T \in \mathcal{K}$, then $TSx = \lambda(S)Tx = \lambda(S)\lambda(T)x$ and $T \mapsto \lambda(T)$ is continuous. Thus every such eigenvalue function λ is a character on $\mathcal{K}$.

1.5   Operators with relatively weakly compact orbits

We now apply the abstract setting to discrete operator semigroups.

Definition 1.13. An operator T on a Banach space X has *relatively weakly compact orbits* if the set $\mathcal{T} := \{T^n : n \in \mathbb{N}_0\}$ satisfies one of the equivalent conditions in Lemma 1.6.

Example 1.14. By the Banach–Alaoglu theorem, every power bounded operator on a reflexive Banach space has relatively weakly compact orbits.

    We now apply the abstract decomposition from Subsection 1.4 to such operators.

Theorem 1.15 (Jacobs–Glicksberg–de Leeuw decomposition). *Let X be a Banach space and let $T \in \mathcal{L}(X)$ have relatively weakly compact orbits. Then $X = X_r \oplus X_s$, where*

$$X_r := \overline{\lin}\{x \in X : Tx = \gamma x \text{ for some } \gamma \in \Gamma\},$$
$$X_s := \{x \in X : 0 \text{ is a weak accumulation point of } \{T^n x : n = 0, 1, 2, \ldots\}\}.$$

    We will formulate an extended version of this theorem later in Theorem II.4.8.

    Finally, we state the Jacobs–Glicksberg–de Leeuw decomposition theorem for the strong operator topology.

Theorem 1.16. *Let X be a Banach space and $T \in \mathcal{L}(X)$ such that for every $x \in X$ the orbit $\{T^n x : n = 0, 1, 2 \ldots\}$ is relatively compact in X. Then $X = X_r \oplus X_s$ for*

$$X_r := \overline{\lin}\{x \in X : Tx = \gamma x \text{ for some } \gamma \in \Gamma\},$$
$$X_s := \{x \in X : \|T^n x\| \to 0 \text{ as } n \to \infty\}.$$

    The proof follows directly from Theorem 1.15, the fact that on compact sets weak convergence implies convergence, and since $\lim_{j \to \infty} \|T^{n_j} x\| = 0$ for some subsequence $\{n_j\}$ implies $\lim_{n \to \infty} \|T^n x\| = 0$, see Lemma II.2.4.

1.6   Jacobs–Glicksberg–de Leeuw decomposition for C_0-semigroups

We now apply the abstract setting of Subsection 1.4 to C_0-semigroups. Note that the results of this section are completely analogous to the discrete version considered in Subsection 1.5.

Definition 1.17. A C_0-semigroup $T(\cdot)$ on a Banach space X is called *relatively weakly compact* if the set $\mathcal{T} := \{T(t) : t \geq 0\} \subset \mathcal{L}(X)$ satisfies one of the equivalent conditions in Lemma 1.6.

Remark 1.18. A C_0-semigroup $T(\cdot)$ on a Banach space X is relatively weakly compact if and only if $T(1)$ has relatively weakly compact orbits. The non-trivial "if" implication follows from the semigroup law and the fact that the sets $\{T(r)x : r \in [0, 1]\}$ are (strongly) compact as the continuous image of $[0, 1]$.

The following theorem is again a special case of the abstract Jacobs–Glicksberg–de Leeuw decomposition and is fundamental for the stability theory of C_0-semigroups, see Engel, Nagel [78], Theorem V.2.8.

Theorem 1.19 (Jacobs–Glicksberg–de Leeuw decomposition for C_0-semigroups).
Let X be a Banach space and $T(\cdot)$ be a relatively weakly compact C_0-semigroup on X. Then $X = X_r \oplus X_s$, where

$$X_r := \overline{\lin}\{x \in X : T(t)x = e^{i\alpha t}x \text{ for some } \alpha \in \mathbb{R} \text{ and all } t \geq 0\},$$

$$X_s := \{x \in X : 0 \text{ is a weak accumulation point of } \{T(t)x : t \geq 0\}\}.$$

In Theorem III.5.7 we formulate an extended version of the above theorem. Moreover, we refer to Arendt, Batty, Hieber, Neubrander [10, Theorem 5.4.11] for an individual version.

We now state the decomposition theorem of Jacobs–Glicksberg–de Leeuw for C_0-semigroups with respect to the strong operator topology.

Theorem 1.20. *Let X be a Banach space and $T(\cdot)$ be relatively compact in the strong operator topology, i.e., for every $x \in X$ the orbit $\{T(t)x : t \geq 0\}$ is relatively compact in X. Then $X = X_r \oplus X_s$ for*

$$X_r := \overline{\lin}\{x \in X : T(t)x = e^{i\alpha t}x \text{ for some } \alpha \in \mathbb{R} \text{ and all } t \geq 0\},$$
$$X_s := \{x \in X : \|T(t)x\| \to 0 \text{ as } t \to \infty\}.$$

The proof follows as in the discrete case directly from Theorem 1.19, the fact that, for sequences in a compact set, weak convergence implies convergence, and Lemma III.3.4.

For an individual version of Theorem 1.20 see Arendt, Batty, Hieber, Neubrander [10, Theorem 5.4.6].

Remark 1.21. Theorem 1.20 applies, for instance, to eventually compact C_0-semigroups or bounded C_0-semigroups whose generator has compact resolvent, see Engel, Nagel [78, Corollary V.2.15].

2 Mean ergodicity

We now introduce mean ergodicity, or convergence in mean, of operators and C_0-semigroups. This is an important asymptotic property having its origin in the classical "mean" and "individual ergodic theorems" by von Neumann and Birkhoff (see books on ergodic theory such as Halmos [118], Petersen [212], Walters [254]) which occurs in quite general situations.

2.1 Mean ergodic operators

We first look at mean ergodicity of operators and follow Yosida [260, Section VIII.3], Nagel [195] as well as Engel, Nagel [78, Section V.4] where the continuous case is treated.

We begin with the so-called Cesàro means of an operator.

Definition 2.1. For a bounded operator T on a Banach space X the *Cesàro means* S_n are defined by

$$S_n x := \frac{1}{n+1} \sum_{k=0}^{n} T^k x, \quad x \in X.$$

We are interested in convergence of S_n. The following easy lemma already gives some information.

Lemma 2.2. *Let T be a bounded operator on a Banach space X. If T satisfies* $\lim_{n \to \infty} \frac{\|T^n x\|}{n} = 0$ *for every $x \in X$, then $S_n x$ converges as $n \to \infty$ for every*

$$x \in \operatorname{Fix} T \oplus \operatorname{rg}(I - T).$$

More precisely, $S_n x = x$ for every $x \in \operatorname{Fix} T$ and $\lim_{n \to \infty} S_n x = 0$ for every $x \in \operatorname{rg}(I - T)$.

Proof. To prove the second statement take $x = z - Tz \in \operatorname{rg}(I - T)$. Then we have

$$S_n x = \frac{1}{n+1} \sum_{k=0}^{n} \left(T^k z - T^{k+1} z \right) = \frac{z - T^{n+1} z}{n+1} \to 0 \quad \text{as } n \to \infty$$

by assumption.   □

Definition 2.3. A bounded operator T on a Banach space X is called *mean ergodic* if the Cesàro means $\{S_n x\}_{n=0}^{\infty}$ converge as $n \to \infty$ for every $x \in X$. In this case the limit

$$x \mapsto Px := \lim_{n \to \infty} S_n x$$

is called the *mean ergodic projection* corresponding to T.

Remark 2.4.   1) For a mean ergodic operator T, the operator P is indeed a projection commuting with T since $P = TP = PT$ follows from the definition and implies $P = T^n P = \lim_{n \to \infty} \frac{1}{n+1} \sum_{k=0}^{n} T^k P = P^2$. If T is a contraction on a Hilbert space, then P is an orthogonal projection since it is contractive.

2) By $\frac{T^n x}{n} = S_n x - \frac{n-1}{n} S_{n-1} x$, every mean ergodic operator T satisfies

$$\lim_{n \to \infty} \frac{\|T^n x\|}{n} = 0 \quad \text{for every } x \in X,$$

and hence $r(T) \leq 1$ by the uniform boundedness principle.

As a counterpart to the Cesàro means we now introduce Abel means of an operator.

Definition 2.5. For a bounded operator T on a Banach space X, the *Abel means* of T are the operators $\tilde{S}_r$ defined by

$$\tilde{S}_r x := (r - 1) \sum_{n=0}^{\infty} \frac{T^n x}{r^{n+1}}, \quad x \in X$$

for $r > 1$, whenever the above series converges.

The following classical property is central for our study, see, e.g., Hardy [123, Section V.12], Emilion [76] or Shaw [230] for the continuous case.

Lemma 2.6 (Equivalence of the Cesàro and Abel means). *Let $\{a_n\}_{n=1}^{\infty}$ be a sequence in a Banach space X such that the Abel means $(r - 1) \sum_{n=0}^{\infty} \frac{a_n}{r^{n+1}}$ exist for every $r > 1$. Then convergence of the Cesàro means as $n \to \infty$ implies convergence of the Abel means as $r \to 1+$ and the limits coincide, i.e.,*

$$\lim_{n \to \infty} \frac{1}{n+1} \sum_{k=0}^{n} a_k = \lim_{r \to 1+} (r - 1) \sum_{n=0}^{\infty} \frac{a_n}{r^{n+1}}.$$

Conversely, convergence of the Abel means implies convergence of the Cesàro means and the limits coincide in each of the following cases:

- *$\{a_n\}_{n=1}^{\infty}$ is bounded;*
- *$X = \mathbb{C}$ and $a_n \geq 0$ for every $n \in \mathbb{N}$.*

In particular, for a bounded operator T on a Banach space with $r(T) \leq 1$, one has

$$\lim_{n \to \infty} \frac{1}{n+1} \sum_{k=0}^{n} T^k = \lim_{r \to 1+} (r - 1) R(r, T)$$

in the weak, strong and norm operator topology, whenever the left limit exists.

Note that the last assertion follows from the previous ones by the Neumann series representation of the resolvent.

Remark 2.7. Shaw [230] showed that convergence of Abel means implies convergence of the Cesàro means and the limits coincide for every positive sequence in a Banach lattice.

We now state an easy but useful property of mean ergodic operators. We will see in Theorem 2.9 that it is in fact equivalent to mean ergodicity.

Proposition 2.8. *Let T be a mean ergodic operator on a Banach space X. Then the mean ergodic projection P yields a decomposition*

$$X = \operatorname{rg} P \oplus \ker P$$

with

$$\operatorname{rg} P = \operatorname{Fix} T, \quad \ker P = \overline{\operatorname{rg}(I - T)}.$$

Moreover, the projection P can also be obtained as

$$Px = \lim_{r \to 1+} (r - 1)R(r, T)x \quad \text{for all } x \in X. \tag{I.3}$$

Proof. The inclusion $\operatorname{Fix} T \subset \operatorname{rg} P$ follows from the definition of P. For the converse inclusion take some $z = Px \in \operatorname{rg} P$. Then $Tz = TPx = Px = z$ by the above remark.

The inclusion $\overline{\operatorname{rg}(I - T)} \subset \ker P$ follows from Lemma 2.2. For the converse inclusion take $y \in X'$ vanishing on $\overline{\operatorname{rg}(I - T)}$. This means that $\langle x, y \rangle = \langle Tx, y \rangle$ for every $x \in X$, which implies $\langle x, y \rangle = \langle Px, y \rangle$ for every $x \in X$, hence y vanishes on $\ker P$, and we obtain $\ker P \subset \overline{\operatorname{rg}(I - T)}$.

The last formula follows from $r(T) \leq 1$ and Lemma 2.6. $\qquad\qquad\square$

The following classical theorem gives different characterisations of mean ergodicity, see, e.g., Yosida [260, Theorem VIII.3.2], Nagel [195] and Krengel [154, § 2.1].

Theorem 2.9 (Mean ergodic theorem). *Let T be a bounded operator on a Banach space X satisfying $\sup_{n \in \mathbb{N}} \|T^n\| < \infty$. Then the following assertions are equivalent.*

(i) *T is mean ergodic.*

(ii) *For every $x \in X$ there exists a subsequence $\{n_k\}_{k=1}^{\infty}$ such that $S_{n_k} x$ converges weakly as $k \to \infty$.*

(iii) *$\lim_{r \to 1+} (r - 1)R(r, T)x$ exists for every $x \in X$.*

(iv) *$\operatorname{Fix} T$ separates $\operatorname{Fix}(T')$.*

(v) *$X = \operatorname{Fix} T \oplus \overline{\operatorname{rg}(I - T)}$.*

In particular, operators with relatively weakly compact orbits are mean ergodic.

Proof. The equivalence (i)$\Leftrightarrow$(iii) follows from Lemma 2.6.

The implication (i)$\Rightarrow$(ii) is trivial.

For the implication (ii)$\Rightarrow$(iv) take $0 \neq y \in \operatorname{Fix} T'$. We have to find $x \in \operatorname{Fix} T$ with $\langle x, y \rangle \neq 0$. Take $x_0 \in X$ with $\langle x_0, y \rangle \neq 0$. By (ii) there exists weak-$\lim_{k \to \infty} S_{n_k} x_0 =: x$ for some subsequence $\{n_k\}_{k=0}^{\infty}$. We now show that $x \in \operatorname{Fix} T$. Observe

$$x - Tx = (I - T)(x - S_{n_k} x_0) + (I - T)S_{n_k} x_0 = (I - T)(x - S_{n_k} x_0) + \frac{x_0 - T^{n_k + 1} x_0}{n_k + 1}.$$

Since T is power bounded, the second summand on the right-hand side tends to 0 as $k \to \infty$. Thus weak continuity of $(I - T)$ implies

$$x - Tx = \text{weak-}\lim_{k \to \infty} (I - T)(x - S_{n_k} x_0) = 0,$$

hence $x \in \operatorname{Fix} T$. Therefore we obtain

$$\langle x, y \rangle = \lim_{k \to \infty} \langle S_{n_k} x_0, y \rangle = \langle x_0, y \rangle \neq 0,$$

where we have used $y \in \ker T'$, and (iv) follows.

Assume now that (iv) holds. For (v) it suffices to show that the subspace

$$G := \operatorname{Fix} T \oplus \operatorname{rg}(I - T)$$

is dense in X. Take $y \in X'$ vanishing on G. Then $\langle x, y \rangle = \langle Tx, y \rangle$ for every $x \in X$, and hence $y \in \operatorname{Fix} T'$. Since y vanishes on $\operatorname{Fix} T$ by assumption as well, we have by (iv) that $y = 0$, and density of G follows.

To prove (v)$\Rightarrow$(i) we first observe that $S_n x$ converges for every $x \in \operatorname{Fix} T \oplus \operatorname{rg}(I - T)$ by Lemma 2.2. By density of this set and boundedness of $\{S_n\}_{n=0}^{\infty}$ we obtain that S_n converges on the whole X, i.e., T is mean ergodic.

In order to show the last assertion, suppose that T has relatively weakly compact orbits and check assertion (ii). Take $x \in X$ and observe that $\{S_n x : n \in \mathbb{N}\} \subset \operatorname{co}\{T^n x : n = 0, 1, 2, \ldots\}$ and hence is relatively weakly compact by assumption and the Kreĭn–Šmulian theorem. By the Eberlein–Šmulian theorem, weak compactness coincides with weak sequential compactness, and assertion (ii) of Theorem 2.9 follows. $\qquad\square$

The following result shows the value of the various characterisations in the above mean ergodic theorem.

Corollary 2.10. *A power bounded operator T on a Banach space is mean ergodic with mean ergodic projection $P = 0$ if and only if $\operatorname{Fix} T' = \{0\}$.*

The proof uses (i)$\Leftrightarrow$(iv) in Theorem 2.9 and the fact that $\operatorname{Fix} T'$ separates $\operatorname{Fix} T$ for power bounded operators. For a quite large class of operators, mean ergodicity holds automatically as a consequence of the Banach–Alaoglu theorem and Theorem 2.9.

Corollary 2.11. *Every power bounded operator on a reflexive Banach space is mean ergodic.*

Note that mean ergodicity of power bounded operators characterises reflexivity for Banach spaces with unconditional basis, see Fonf, Lin, Wojtaszczyk [89] and also the discussion in Emel'yanov [75, Section 2.2.3].

Remark 2.12. If $P_\sigma(T) \cap \Gamma \subset \{1\}$ and T has relatively weakly compact orbits, then the mean ergodic projection coincides with the projection from the Jacobs–Glicksberg–de Leeuw decomposition, see Section 1.5.

There are many extensions of the mean ergodic theorem, see e.g. Berend, Lin, Rosenblatt, Tempelman [31] for recent results on modulated and subsequential ergodic theorems and further references.

2.2 Uniformly mean ergodic operators

In this subsection we consider a stronger property than the above mean ergodicity, see also Krengel [154, § 2.2].

Definition 2.13. An operator T on a Banach space is called *uniformly mean ergodic* if the Cesàro means S_n converge in the norm operator topology.

In other words, an operator T is uniformly mean ergodic if and only if T is mean ergodic and one has

$$\lim_{n \to \infty} \|S_n - P\| = 0$$

for the mean ergodic projection P.

Remark 2.14. A uniformly mean ergodic operator T again satisfies $r(T) \le 1$, and the projection P can be obtained as

$$P = \lim_{r \to 1+} (r - 1)R(r, T)$$

by Lemma 2.6. Recall that for a mean ergodic operator this formula holds only pointwise.

The following result characterises uniform mean ergodicity, see Lin [168].

Theorem 2.15. *Let T be a power bounded operator on a Banach space. Then the following assertions are equivalent.*

(i) *T is uniformly mean ergodic.*

(ii) *There exists $\lim_{r \to 1+}(r - 1)R(r, T)$.*

(iii) *$1 \in \rho(T)$ or 1 is a first-order pole of $R(\lambda, T)$.*

(iv) *$\mathrm{rg}(I - T)$ is closed in X.*

(v) *$X = \mathrm{Fix}\, T \oplus \mathrm{rg}(I - T)$.*

Proof. The equivalence (i)⇔(ii) again follows from Lemma 2.6.

(i)⇒(iii). Assume that $1 \in \sigma(T)$. We have to show that 1 is a first-order pole of the resolvent. By Proposition 2.8, the decomposition into invariant subspaces $X = \mathrm{Fix}\, T \oplus \overline{\mathrm{rg}(I - T)}$ holds. Since $T|_{\mathrm{Fix}\, T} = I$, and hence 1 is a first-order pole of $R(\lambda, T|_{\mathrm{Fix}\, T})$, it suffices to show that

$$1 \in \rho(T|_{\ker P}). \tag{I.4}$$

Assume the opposite, i.e., that $1 \in \sigma(T|_{\ker P})$. Since T is power bounded, we have $1 \in A_\sigma(T|_{\ker P})$. Thus there exists an approximate eigenvalue $\{x_m\}_{m=1}^{\infty} \subset \ker P$, $\|x_m\| = 1$, such that $\lim_{m \to \infty} \|(I - T)x_m\| = 0$. Then we obtain

$$\lim_{m \to \infty} (I - T^n)x_m = \lim_{m \to \infty} \sum_{k=0}^{n-1} T^k(I - T)x_m = 0$$

for every $n \in \mathbb{N}$. This implies $\lim_{m \to \infty}(I - S_n)x_m = 0$, so $1 \in A_\sigma(S_n)$ and hence $\|S_n|_{\ker P}\| \geq 1$ for every $n \in \mathbb{N}$. This contradicts the uniform ergodicity of T which means $\lim_{n \to \infty} \|S_n|_{\ker P}\| = 0$. Hence (I.4) is proved.

The implication (iii)$\Rightarrow$(ii) is clear.

(ii)$\Rightarrow$(iv). Let $P := \lim_{r \to 1+}(r - 1)R(r, T)$. We first prove that $Px = 0$ for every $x \in \overline{\mathrm{rg}(I - T)}$. Take $x = z - Tz \in \mathrm{rg}(I - T)$. Then we have, by the resolvent identity,

$$(r - 1)R(r, T)x = (r - 1)R(r, T)(I - T)z = (r - 1)z - (r - 1)^2 R(r, T)z$$

and hence by (iii) $Px = \lim_{r \to 1+}(r - 1)R(r, T)x = 0$. So $P = 0$ on $Y := \overline{\mathrm{rg}(I - T)}$, i.e.,

$$\lim_{r \to 1+} \|(r - 1)R(r, T)|_Y\| = 0$$

holds. (We used here that Y is $R(r, T)$-invariant by the Neumann representation of the resolvent.) This implies that the operator

$$(I - T)R(r, T) = I - (r - 1)R(r, T)$$

is invertible on Y for every $r > 1$ small enough. So we have in particular

$$Y = \overline{\mathrm{rg}(I - T)} = (I - T)R(r, T)Y \subset \mathrm{rg}(I - T),$$

and $\mathrm{rg}(I - T)$ is closed.

(iv)$\Rightarrow$(v). We first show that $(I - T)|_{\mathrm{rg}(I-T)}$ is invertible. Assume the opposite, i.e., that $1 \in \sigma(T|_{\mathrm{rg}(I-T)})$. As in the proof of (i)$\Rightarrow$(iii), this implies that $1 \in \sigma(S_n|_{\mathrm{rg}(I-T)})$ and hence $\|S_n|_{\mathrm{rg}(I-T)}\| \geq 1$ for every $n \in \mathbb{N}$. On the other hand, for $x = z - Tz \in \mathrm{rg}(I - T)$ we have, by

$$(n + 1)S_n x = \sum_{k=0}^{n}(T^k z - T^{k+1}z) = z - T^{n+1}z$$

and power boundedness of T, that $\sup_{n \in \mathbb{N}} \|nS_n x\| < \infty$. By closedness of $\mathrm{rg}(I-T)$ and the uniform boundedness principle this implies $\sup_{n \in \mathbb{N}} \|nS_n|_{\mathrm{rg}(I-T)}\| < \infty$ and hence $\lim_{n \to \infty} \|S_n|_{\mathrm{rg}(I-T)}\| = 0$, a contradiction. So $(I - T)|_{\mathrm{rg}(I-T)}$ is invertible.

Take now $x \in X$. By invertibility of $(I - T)|_{\mathrm{rg}(I-T)}$ there exists a unique $z \in \mathrm{rg}(I-T)$ such that $(I-T)x = (I-T)z$. Then $(x-z) \in \mathrm{Fix}\, T$ and $x = (x-z)+z$, and the decomposition $X = \mathrm{Fix}\, T \oplus \mathrm{rg}(I - T)$ is proved.

(v)$\Rightarrow$(i). Observe that $\mathrm{Fix}\, T$ and $\mathrm{rg}(I - T)$ are invariant closed subspaces. On $\mathrm{Fix}\, T$ the restriction of S_n is equal to I, so $T|_{\mathrm{Fix}\, T}$ is uniformly mean ergodic. Moreover, as in the proof of the implication (iv)$\Rightarrow$(v) we have (by closedness of $\mathrm{rg}(I - T)$)

$$\lim_{n \to \infty} \|S_n|_{\mathrm{rg}(I-T)}\| = 0.$$

So T is uniformly mean ergodic by (v). $\square$

A direct corollary of the above characterisation is the following.

Corollary 2.16. *A quasi-compact power bounded operator T on a Banach space is uniformly mean ergodic.*

Proof. Since T is quasi-compact, we have a decomposition into invariant subspaces $X = X_1 \oplus X_2$ and $T = T_1 \oplus T_2$ such that $\dim X_1 < \infty$ and $r(T_2) < 1$. Assume that $1 \in \sigma(T)$ which is equivalent to $1 \in \sigma(T_1)$. Since T_1 is a power bounded matrix, we see by Jordan's representation that 1 is the first-order pole of $R(\lambda, T_1)$. Thus T satisfies condition (iii) of Theorem 2.15.   □

Remark 2.17. For positive operators on Banach lattices a stronger assertion holds: A positive operator T is quasi-compact if and only if it is mean ergodic and satisfies $\dim \operatorname{Fix} T < \infty$, see Lin [170].

2.3 Mean ergodic C_0-semigroups

We now characterise mean ergodicity of C_0-semigroups. Most of the proofs are analogous to the discrete case, so we omit them and refer to Engel, Nagel [78, Section V.4] for details.

We first define Cesàro means of a C_0-semigroup $T(\cdot)$.

Definition 2.18. The *Cesàro means* of a C_0-semigroup $T(\cdot)$ are the operators

$$S(t)x := \frac{1}{t} \int_0^t T(s)x\,ds, \quad x \in X \text{ and } t > 0.$$

The following easy lemma describes convergence of the Cesàro means on a subspace. Here and later we use the notation

$$\operatorname{Fix} T(\cdot) := \bigcap_{t \geq 0} \operatorname{Fix} T(t).$$

Lemma 2.19. *Let $T(\cdot)$ be a C_0-semigroup on a Banach space X. If T satisfies $\lim_{t \to \infty} \frac{\|T(t)x\|}{t} = 0$ for every $x \in X$, then $S(t)x$ converges as $t \to \infty$ for every*

$$x \in \operatorname{Fix} T(\cdot) \oplus \operatorname{lin} \cup_{t \geq 0} \operatorname{rg}(I - T(t)).$$

More precisely, $S(t)x = x$ for every $x \in \operatorname{Fix} T(\cdot)$ and $\lim_{t \to \infty} S(t)x = 0$ for every $x \in \operatorname{lin} \cup_{t \geq 0} \operatorname{rg}(I - T(t))$.

Mean ergodic semigroups are defined as follows.

Definition 2.20. A C_0-semigroup $T(\cdot)$ on a Banach space X is called *mean ergodic* if the Cesàro means $S(t)x$ converge as $t \to \infty$ for every $x \in X$. In this case the limit

$$x \mapsto Px := \lim_{t \to \infty} S(t)x$$

is called the *mean ergodic projection* corresponding to $T(\cdot)$.

Remark 2.21. Mean ergodicity of $T(\cdot)$ implies $P = T(t)P = PT(t)$ and therefore $Px = T(t)Px = \lim_{t\to\infty} \frac{1}{t}\int_0^t T(s)Pxds = P^2x$. So P is indeed a projection commuting with $T(\cdot)$.

As above, we also consider the Abel means.

Definition 2.22. The *Abel means* of a C_0-semigroup $T(\cdot)$ with $\omega_0(T) \leq 0$ are the operators $\tilde{S}_a$ defined by

$$\tilde{S}_a x := a \int_0^\infty e^{-as}T(s)x\,ds \quad \text{for } x \in X \text{ and } a > 0.$$

The following relation between convergence of the Cesàro and Abel means is essential for the resolvent approach to asymptotics of C_0-semigroups, see, e.g., Emilion [76] and Shaw [230].

Lemma 2.23 (Equivalence of the Cesàro and Abel means, continuous case). *Let X be a Banach space and $f : \mathbb{R}_+ \to X$ be continuous such that the Abel means $a \int_0^\infty e^{-at}f(t)dt$ exist for every $a > 0$. Then convergence of the Cesàro means as $t \to \infty$ implies convergence of the Abel means as $a \to 0+$ and the limits coincide, i.e.,*

$$\lim_{t\to\infty} \frac{1}{t}\int_0^t f(s)ds = \lim_{a\to 0+} a \int_0^\infty e^{-at}f(t)dt.$$

Conversely, convergence of the Abel means implies convergence of the Cesàro means and the limits coincide in each of the following cases:

- f *is bounded;*

- $X = \mathbb{C}$ *and $f(s) \geq 0$ for every $s \geq 0$.*

In particular, for a C_0-semigroup $T(\cdot)$ on a Banach space X with generator A satisfying $\omega_0(T) \leq 0$ one has

$$\lim_{t\to\infty} \frac{1}{t}\int_0^t T(s)xds = \lim_{a\to 0+} aR(a, A)x$$

for every $x \in X$, whenever the left limit exists.

The following proposition gives more information on the decomposition obtained from the mean ergodic projection.

Proposition 2.24. *Let $T(\cdot)$ be a mean ergodic C_0-semigroup on a Banach space X satisfying $\lim_{t\to\infty} \frac{\|T(t)x\|}{t} = 0$ for every $x \in X$. Then the mean ergodic projection P yields the decomposition*

$$X = \operatorname{rg} P \oplus \ker P$$

with

$$\operatorname{rg} P = \ker A = \operatorname{Fix} T(\cdot), \quad \ker P = \overline{\operatorname{rg} A} = \overline{\operatorname{lin}} \cup_{t\geq 0} \operatorname{rg}(I - T(t)).$$

Moreover, the projection P can be obtained as

$$Px = \lim_{a \to 0+} aR(a, A)x \quad \text{for all } x \in X. \tag{I.5}$$

The following classical theorem characterises mean ergodicity in various ways.

Theorem 2.25 (Mean ergodic theorem for C_0-semigroups). *Let $T(\cdot)$ be a bounded C_0-semigroup on a Banach space X. Then the following assertions are equivalent.*

(i) *$T(\cdot)$ is mean ergodic.*

(ii) *For every $x \in X$ there exists a sequence $\{t_k\}_{k=1}^{\infty} \subset \mathbb{R}_+$ converging to ∞ such that $S(t_k)x$ converges weakly as $k \to \infty$.*

(iii) *There exists $\lim_{a \to 0+} aR(a, A)x$ for every $x \in X$.*

(iv) *$\ker A$ separates $\ker A'$.*

(v) *$X = \ker A \oplus \overline{\operatorname{rg} A}$.*

In particular, every relatively weakly compact C_0-semigroup, and hence every bounded semigroup on a reflexive Banach space, is mean ergodic.

As a direct corollary we obtain an easy characterisation of C_0-semigroups whose Cesàro means converge to 0.

Corollary 2.26. *A bounded C_0-semigroup $T(\cdot)$ on a Banach space with generator A is mean ergodic with mean ergodic projection $P = 0$ if and only if $\ker A' = \{0\}$.*

Remark 2.27. If $T(\cdot)$ is a relatively weakly compact semigroup with generator A and $P_\sigma(A) \cap i\mathbb{R} = \{0\}$, then the mean ergodic projection corresponding to $T(\cdot)$ coincides with the projection from the Glicksberg–Jacobs–de Leeuw decomposition, see Section 1.6.

2.4   Uniformly mean ergodic C_0-semigroups

We now introduce uniformly mean ergodic C_0-semigroups. Again, the situation is similar to the discrete case and we refer to Engel, Nagel [78, Section V.4] for the proofs.

Definition 2.28. A C_0-semigroup $T(\cdot)$ on a Banach space is called *uniformly mean ergodic* if the Cesàro means $S(t)$ converge as $t \to \infty$ in the norm operator topology.

Remark 2.29. A C_0-semigroup $T(\cdot)$ is uniformly mean ergodic if and only if it is mean ergodic and

$$\lim_{t \to \infty} \|S(t) - P\| = 0$$

holds for the mean ergodic projection P.

The following theorem characterises uniform mean ergodicity of C_0-semigroups, see Lin [169].

Theorem 2.30. *Let $T(\cdot)$ be a bounded C_0-semigroup on a Banach space. Then the following assertions are equivalent.*

(i) *$T(\cdot)$ is uniformly mean ergodic.*

(ii) *There exists $\lim_{a\to 0+} aR(a,T)$.*

(iii) *$0 \in \rho(A)$ or 0 is a first-order pole of $R(\lambda, A)$.*

(iv) *$\operatorname{rg} A$ is closed in X.*

(v) *$X = \operatorname{Fix} T(\cdot) \oplus \operatorname{rg} A$.*

In this case, the mean ergodic projection satisfies $P = \lim_{a\to 0+} aR(a,T)$.

A direct corollary of the above characterisation is the following example of uniformly mean ergodic semigroups.

Corollary 2.31. *A quasi-compact bounded C_0-semigroup $T(\cdot)$ on a Banach space is uniformly mean ergodic.*

3 Specific concepts from semigroup theory

In this section we present tools and methods from semigroup theory which we will need later. The main reference for basic definitions and facts from semigroup theory is Engel, Nagel [78].

3.1 Cogenerator

Another useful tool, besides the generator, in the theory of C_0-semigroups is the so-called cogenerator. It can be obtained easily from the generator (see formula (I.6) below), it is a bounded operator, and, as we will see later, reflects many properties of the semigroup. The diagram on top of the next page shows the major objects related to a C_0-semigroup.

The cogenerator is defined as follows.

Definition 3.1. Let A generate a C_0-semigroup $T(\cdot)$ on a Banach space X satisfying $1 \in \rho(A)$. The operator V defined by

$$V := (A + I)(A - I)^{-1} \in \mathcal{L}(X)$$

is called the *cogenerator* of $T(\cdot)$.

Remark 3.2. The identity

$$V = (A - I + 2I)(A - I)^{-1} = I - 2R(1, A) \tag{I.6}$$

implies that $V - I$ has a densely defined inverse $(V - I)^{-1} = \frac{1}{2}(A - I)$. In particular,

$$A = (V + I)(V - I)^{-1} = I + 2(V - I)^{-1}$$

holds, i.e., the generator is also the (negative) Cayley transform of the cogenerator.

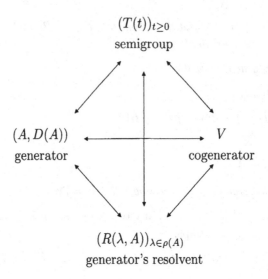

Note that the cogenerator determines the generator, and therefore the semigroup, uniquely. As a further consequence of (I.6) one has the following easy description the spectrum and resolvent of V.

Proposition 3.3. *The spectrum of the cogenerator V is*

$$\sigma(V) \setminus \{1\} = \left\{ \frac{\lambda+1}{\lambda-1} : \lambda \in \sigma(A) \right\}.$$

The same relation holds for the point spectrum, residual point spectrum and approximative point spectrum, respectively. Moreover,

$$R(\lambda, A) = \frac{1}{\lambda-1}(I - V)R\left(\frac{\lambda+1}{\lambda-1}, V \right) \tag{I.7}$$

holds for every $\lambda \in \rho(A) \setminus \{1\}$.

Proof. The assertion on the spectrum of V is a direct consequence of (I.6) and the spectral mapping theorem for the resolvent (see e.g. Engel, Nagel [78, Theorem IV.1.13]), while (I.7) follows from

$$\lambda I - A = (\lambda V - \lambda I - V - I)(V - I)^{-1}$$
$$= (V(\lambda - 1) - (\lambda + 1))(V - I)^{-1} = (\lambda - 1)\left(\frac{\lambda+1}{\lambda-1} - V \right)(I - V)^{-1}$$

for every $\lambda \neq 1$.   □

It is remarkable that for operators on Hilbert spaces there is a very simple characterisation of operators being the cogenerator of a contraction semigroup. This is the following theorem due to Foiaş, Sz.-Nagy [238, Theorem III.8.1]. The proof below of the non-trivial "if"-implication (see also Katz [145]) is simpler than the original one which uses a special functional calculus to construct the semigroup.

Theorem 3.4 (Foiaş–Sz.-Nagy). *Let H be a Hilbert space and $V \in \mathcal{L}(H)$. Then V is the cogenerator of a contractive C_0-semigroup if and only if V is contractive and $1 \notin P_\sigma(V)$.*

Proof. Assume first that V is the cogenerator of a contractive C_0-semigroup $T(\cdot)$ and denote by A its generator. Then the operator $I - V = 2R(1, A)$ is injective and hence $1 \notin P_\sigma(V)$. Moreover, by the Lumer–Phillips Theorem (see, e.g., Engel, Nagel [78, Theorem II.3.15]) A is dissipative and hence

$$\|(A + I)x\|^2 - \|(A - I)x\|^2 = 4\mathrm{Re}\,\langle Ax, x \rangle \leq 0 \qquad \forall x \in D(A).$$

Therefore $\|Vx\| = \|(A + I)(A - I)^{-1}x\| \leq \|(A - I)(A - I)^{-1}x\| = \|x\|$ for every $x \in H$, hence V is contractive.

Let now V be a contraction with $1 \notin P_\sigma(V)$. Then the operator $I - V$ is injective, and we can define

$$A := -(I + V)(I - V)^{-1} \qquad \text{with} \quad D(A) := \mathrm{rg}(I - V).$$

Note that $A = I - 2(I - V)^{-1}$ holds.

We show first that $\mathrm{Re}\,\langle Ax, x \rangle \leq 0$ for every $x \in D(A)$. Indeed, for $x \in D(A)$ and $y := (V - I)^{-1}x$ we have

$$\langle Ax, x \rangle = \langle (I + V)(V - I)^{-1}x, x \rangle = \langle (I + V)y, (V - I)y \rangle$$
$$= \|Vy\|^2 - \|y\|^2 + 2i \cdot \mathrm{Im}\,\langle y, Vy \rangle$$

and therefore $\mathrm{Re}\,\langle Ax, x \rangle \leq 0$.

We observe further that $(I - A)^{-1} = \frac{1}{2}(I - V)$ and therefore $1 \in \rho(A)$. Moreover, since V is mean ergodic, we have $\overline{\mathrm{rg}(I - V)} = H$ by Proposition 2.8, so the operator A is densely defined.

The assertion now follows directly from the Lumer–Phillips Theorem (see, e.g., Engel, Nagel [78, Theorem II.3.15]). $\square$

Further, many properties of a contraction semigroup on a Hilbert space can be seen from its cogenerator. The following is again due to Foiaş, Sz.-Nagy, see [238, Sections III.8–9].

Theorem 3.5. *Let $T(\cdot)$ be a contractive C_0-semigroup on a Hilbert space with cogenerator V. Then $T(\cdot)$ is normal, self-adjoint, isometric or unitary if and only if V is normal, self-adjoint, isometric or unitary, respectively.*

Note that all the equivalences except the isometry property follow easily from the spectral theorem in its multiplicator form, see e.g. Conway [52, Theorem IX.4.6].

The cogenerator approach allows us to transfer many properties of single operators to C_0-semigroups in a short and elegant way. However, it is not yet clear how this approach extends to C_0-semigroups on general Banach spaces.

To conclude this subsection we present an elementary but useful formula for the powers of the cogenerator in terms of the semigroup, see e.g. Gomilko [105].

Lemma 3.6. *Let $T(\cdot)$ be a C_0-semigroup on a Banach space X with $\omega_0(T) < 1$ and cogenerator V. Then*

$$V^n x = x - 2 \int_0^\infty L^1_{n-1}(2t) e^{-t} T(t) x \, dt \qquad (I.8)$$

holds for every $x \in X$, where $L^1_n(t)$ is the first generalized Laguerre polynomial given by

$$L^1_n(t) = \sum_{m=0}^{n} \frac{(-1)^m}{m!} \binom{n+1}{n-m} t^m. \qquad (I.9)$$

Proof. By $V = I - 2R(1, A)$ and the Laplace representation for the resolvent we obtain

$$V^n x = (I - 2R(1, A))^n x = x - \sum_{m=1}^{n} \frac{2^m}{(m-1)!} \binom{n}{n-m} R^{(m-1)}(1, A) x$$

$$= x - \sum_{m=1}^{n} \frac{2^m (-1)^{m-1}}{(m-1)!} \binom{n}{n-m} \int_0^\infty e^{-t} t^{m-1} T(t) x \, dt$$

$$= x - 2 \int_0^\infty L^1_{n-1}(2t) e^{-t} T(t) x \, dt,$$

and the lemma is proved.   □

3.2   Inverse Laplace transform for C_0-semigroups

Our main tool in the resolvent approach to stability of C_0-semigroups is a formula for the inverse Laplace transform of the semigroup.

We first need a lemma on the behaviour of the resolvent of an operator on the left of its abscissa of uniform boundedness.

Lemma 3.7. *Let A be a closed operator with $s_0(A) < \infty$. Then*

$$\|R(z, A)x\| \to 0 \quad \text{as } |z| \to \infty, \ \operatorname{Re} z \geq a \qquad (I.10)$$

holds for every $a > s_0(A)$ and $x \in X$.

Proof. Take any $a > s_0(A)$. Then there exists a constant $M > 0$ such that $\|R(z, A)\| \leq M$ for all $z \in \mathbb{C}$ with $\operatorname{Re} z \geq a$. Take now $x \in D(A)$ and z with $\operatorname{Re} z \geq a$. Then

$$\|R(z, A)x\| = \frac{1}{|z|} \|x + R(z, A)Ax\| \leq \frac{1}{|z|} (\|x\| + M\|Ax\|),$$

and therefore we have

$$\|R(z, A)x\| \to 0 \quad \text{as } |z| \to \infty, \ \text{Re}\, z \ge a.$$

Since $D(A)$ is dense in X and the resolvent of A is uniformly bounded on $\{z : \text{Re}\, z \ge a\}$, this is true for all $x \in X$ and property (I.10) is proved. $\square$

The following representation holds for every C_0-semigroup and is based on arguments from Fourier analysis, see van Neerven [204, Thm.1.3.3] and Kaashoek, Verduyn Lunel [139].

Theorem 3.8. *Let $T(\cdot)$ be a C_0-semigroup on a Banach space X with generator A. Then*

$$T(t)x = \frac{1}{2\pi}(C,1)\int_{-\infty}^{\infty} e^{(a+is)t} R(a+is, A)x\, ds$$

$$= \frac{1}{2\pi t}(C,1)\int_{-\infty}^{\infty} e^{(a+is)t} R^2(a+is, A)x\, ds$$

holds for all $a > s_0(A)$, $t > 0$ and $x \in X$.

Here, $(C,1)\int_{-\infty}^{\infty} f(s)\, ds$ denotes the limit of the Cesàro means of $\int_{-\tau}^{\tau} f(s)ds$ and is defined by

$$(C,1)\int_{-\infty}^{\infty} f(s)\, ds := \lim_{N\to\infty} \frac{1}{N}\int_0^N \int_{-\tau}^{\tau} f(s)\, ds\, d\tau,$$

whenever the limit on the right-hand side exists. Recall that Cesàro convergence of an integral is weaker than the existence of its principal value, and the latter is weaker than convergence.

Proof. The second equality follows from integration by parts. To prove the first one take $x \in X$ and $a > \omega_0(T)$. By the representation

$$R(a+is, A)x = \int_0^{\infty} e^{-ist} e^{-at} T(t)x\, dt$$

and the Cesàro convergence of the inverse Fourier transform (see, e.g., Zaanen [261, Theorem 9.1] or Katznelson [146, Theorem VI.1.11]) we obtain

$$e^{-at}T(t)x = \lim_{N\to\infty} \frac{1}{2\pi N}\int_0^N \int_{-\tau}^{\tau} e^{ist} R(a+is, A)x\, ds\, d\tau,$$

and the first formula for $T(\cdot)$ follows for all $a > \omega_0(T)$.

Take now an arbitrary $a > s_0(A)$ and $a_1 > \max\{a, \omega_0(T)\}$. By the previous considerations and Cauchy's theorem it suffices to show that

$$\lim_{N\to\infty} \int_a^{a_1} e^{b\pm iN} R(b \pm iN, A)x\, db = 0$$

for every $x \in X$. However, this follows directly from Lemma 3.7. $\square$

We now consider a linear densely defined operator A satisfying $s_0(A) < \infty$ for which it is not known whether it is a generator or not. The basis of our approach is the following condition:

$$\langle R(a + i\cdot, A)^2 x, y \rangle \in L^1(\mathbb{R}) \quad \text{for all } x \in X, \ y \in X', \tag{I.11}$$

where $a > s_0(A)$. Indeed, this property allows us to construct the inverse Laplace transform of the resolvent of the operator A which actually yields a semigroup which is strongly continuous on $(0, \infty)$. (Note that strong continuity on $(0, \infty)$ does not imply strong continuity in general, see Kaiser, Weis [140].) The result is based on Shi, Feng [231] and Kaiser, Weis [140], see Eisner [64].

Theorem 3.9 (Laplace inversion formula). *Let A be a densely defined linear operator on a Banach space X satisfying $s_0(A) < \infty$ and assume that condition (I.11) holds for all $a > s_0(A)$. Then the bounded linear operators defined by $T(0) = I$ and*

$$T(t)x := \frac{1}{2\pi} \int_{-\infty}^{\infty} e^{(a+is)t} R(a + is, A)x \, ds \tag{I.12}$$

$$= \frac{1}{2\pi t} \int_{-\infty}^{\infty} e^{(a+is)t} R(a + is, A)^2 x \, ds \qquad \text{for all } t > 0, \tag{I.13}$$

where the improper integrals converge in norm, are independent of $a > s_0(A)$. In addition, the family $(T(t))_{t \geq 0}$ is a semigroup which is strongly continuous on $(0, \infty)$ and satisfies

$$\lim_{t \to 0+} T(t)x = x \quad \text{for all } x \in D(A^2). \tag{I.14}$$

Finally, we have

$$R(z, A)x = \int_0^{\infty} e^{-zt} T(t)x \, ds \qquad \text{for all } x \in D(A), \ \operatorname{Re} z > s_0(A). \tag{I.15}$$

Proof. Define $T(0) := I$ and

$$T(t)x := \frac{1}{2\pi} \int_{-\infty}^{\infty} e^{(a+is)t} R(a + is, A)x \, ds \tag{I.16}$$

for all $x \in X$, $t > 0$ and some $a > 0$. We prove first that the integral on the right-hand side of (I.16) converges for all $a > 0$ and all $x \in X$ and does not depend on $a > 0$. For a fixed $t > 0$ and using $\frac{d}{dz} R(z, A) = -R(z, A)^2$, we obtain for any $r > 0$,

$$it \int_{-r}^{r} e^{(a+is)t} R(a + is, A)x \, ds = e^{(a+ir)t} R(a + ir, A)x - e^{(a-ir)t} R(a - ir, A)x$$

$$+ i \int_{-r}^{r} e^{(a+is)t} R(a + is, A)^2 x \, ds.$$

By (I.10), the first two summands converge to zero if $r \to +\infty$. Therefore

$$t \int_{-\infty}^{\infty} e^{(a+is)t} R(a+is, A)x \, ds = \int_{-\infty}^{\infty} e^{(a+is)t} R(a+is, A)^2 x \, ds, \qquad (\text{I.17})$$

and by condition (I.11) the integral on the right-hand side converges. Indeed, for all $r, R \in \mathbb{R}$, all $x \in X$, and for $B^* = \{y \in X' : \|y\| = 1\}$ we have, by the uniform boundedness principle, that

$$\left\| \int_r^R e^{ist} R(a+is, A)^2 x \, ds \right\| = \sup_{y \in B^*} \int_r^R \langle e^{ist} R(a+is, A)^2 x, y \rangle \, ds$$

$$\leq \sup_{y \in B^*} \| \langle R(a + i\cdot, A)^2 x, y \rangle \|_1 \leq L_1(a) \|x\|$$

holds for some constant $L_1(a)$ not depending on x. This implies the convergence of the integral on the right-hand side of (I.17).

Therefore the integral on the right-hand side of (I.16) converges and

$$T(t)x = \frac{1}{2\pi t} \int_{-\infty}^{\infty} e^{(a+is)t} R(a+is, A)^2 x \, ds \qquad (\text{I.18})$$

for every $x \in X$ and $t > 0$. We show next that $T(t)$ does not depend on $a > 0$. Indeed, by Cauchy's theorem we obtain

$$\int_{-r}^{r} e^{(a+is)t} R(a+is, A)^2 x \, ds - \int_{-r}^{r} e^{(b+is)t} R(b+is, A)^2 x \, ds$$

$$= -\int_a^b e^{\tau + ir} R(\tau + ir, A)^2 x \, d\tau + \int_a^b e^{\tau - ir} R(\tau - ir, A)^2 x \, d\tau$$

for all $a, b > s_0(A)$. By (I.10) and since $a, b > s_0(A)$, the right-hand side converges to zero as $r \to +\infty$. So we have proved that $T(t)$ does not depend on $a > 0$ and formula (I.18) holds.

From (I.18) we obtain

$$|\langle T(t)x, y \rangle| \leq \frac{e^{at}}{2\pi t} \| \langle R(a + i\cdot, A)^2 x, y \rangle \|_1 \qquad (\text{I.19})$$

and, by the uniform boundedness principle, each $T(t)$ is a bounded linear operator satisfying

$$\|T(t)\| \leq \frac{C e^{at}}{t}, \quad t > 0, \qquad (\text{I.20})$$

for some constant C depending on $a > s_0(A)$.

We now show that (I.15) holds for all $x \in D(A)$. Take $x \in D(A)$, $\operatorname{Re} z > s_0(A)$ and $a \in (s_0(A), \operatorname{Re} z)$. Then, by Fubini's theorem and Cauchy's integral theorem for bounded analytic functions on a right half-plane, we have

$$
\int_0^\infty e^{-zt} T(t)x \, dt = \frac{1}{2\pi} \int_0^\infty e^{-zt} \int_{-\infty}^\infty e^{(a+is)t} R(a+is, A)x \, ds dt
$$

$$
= \frac{1}{2\pi} \int_{-\infty}^\infty \left\{ \int_0^\infty e^{(a+is-z)t} dt \right\} \frac{R(a+is, A)Ax + x}{a+is} \, ds
$$

$$
= \frac{1}{2\pi} \int_{-\infty}^\infty \frac{R(a+is, A)Ax + x}{(a+is)(z-a-is)} \, ds = \frac{R(z, A)Ax + x}{z} = R(z, A)x,
$$

and (I.15) is proved.

We next show strong continuity of our semigroup on $(0, \infty)$. Since by (I.20) the semigroup is uniformly bounded on all compact intervals in $(0, \infty)$, it is enough to show that (I.14) holds for all $x \in D(A^2)$. (We used here that $D(A^2)$ is dense, see e.g. Engel, Nagel [78, pp. 53–54]).) Take $x \in D(A^2)$, $a > 0$, and observe that

$$
T(t)x - x = \frac{1}{2\pi} \int_{-\infty}^\infty e^{(a+is)t} \left(R(a+is, A)x - \frac{x}{a+is} \right) ds
$$

$$
= \frac{1}{2\pi} \int_{-\infty}^\infty e^{(a+is)t} \frac{R(a+is, A)Ax}{a+is} \, ds.
$$

Moreover, $\|R(a+is, A)Ax\| \leq \frac{c\|A^2 x\|}{1+|a+is|}$ for some constant c. Therefore, by Lebesgue's theorem,

$$
\lim_{t \to 0+} (T(t)x - x) = \frac{1}{2\pi} \int_{-\infty}^\infty \frac{R(a+is, A)Ax}{a+is} \, ds \tag{I.21}
$$

and the integral on the right-hand side converges absolutely. We now show

$$
\int_{-\infty}^\infty \frac{R(a+is, A)Ax}{a+is} \, ds = 0. \tag{I.22}
$$

Again by Cauchy's theorem and (I.10) we have

$$
\left\| \int_{-r}^r \frac{R(a+is, A)Ax}{a+is} \, ds \right\| = \left\| \int_{-\pi/2}^{\pi/2} \frac{ire^{i\varphi}}{a+re^{i\varphi}} R(a+re^{i\varphi}, A)Ax \, d\varphi \right\|
$$

$$
\leq \int_{-\pi/2}^{\pi/2} \|R(a+re^{i\varphi}, A)Ax\| \, d\varphi \to 0, \quad r \to \infty.
$$

So equality (I.22) is proved, and (I.21) implies (I.14) and the strong continuity of our semigroup on $(0, \infty)$.

We finally prove the semigroup law $T(t+s) = T(t)T(s)$. Take $t, s > 0$, $x \in D(A^2)$, and $0 < \lambda < \mu$. Then $R(\lambda, A)x \in D(A)$ and we have by (I.15)

$$
R(\mu, A)R(\lambda, A)x = \int_0^\infty e^{-\mu t} \left\{ \int_0^\infty e^{-\lambda s} T(t)T(s)x \, ds \right\} dt.
$$

On the other hand observe that

$$\frac{R(\lambda, A)x - R(\mu, A)x}{\mu - \lambda} = \int_0^\infty e^{(\lambda-\mu)t} R(\lambda, A)x \, dt - \frac{1}{\mu - \lambda} \int_0^\infty e^{(\lambda-\mu)t} e^{-\lambda t} T(t)x \, dt$$

$$= \int_0^\infty e^{(\lambda-\mu)t} \left\{ \int_0^\infty e^{-\lambda s} T(s)x \, ds \right\} dt - \int_0^\infty e^{(\lambda-\mu)t} \left\{ \int_0^t e^{-\lambda s} T(s)x \, ds \right\} dt$$

$$= \int_0^\infty e^{(\lambda-\mu)t} \left\{ \int_t^\infty e^{-\lambda s} T(s)x \, ds \right\} dt = \int_0^\infty e^{-\mu t} \left\{ \int_t^\infty e^{\lambda(t-s)} T(s)x \, ds \right\} dt$$

$$= \int_0^\infty e^{-\mu t} \left\{ \int_0^\infty e^{-\lambda s} T(s+t)x \, ds \right\} dt.$$

The uniqueness of the Laplace transform implies $T(t+s)x = T(t)T(s)x$ for almost all $t, s > 0$. By strong continuity of $T(\cdot)$ on $(0, \infty)$ we have $T(t+s)x = T(t)T(s)x$ for all $t, s > 0$ and $x \in D(A^2)$. Since $D(A^2)$ is dense and the case $t = 0$ or $s = 0$ is trivial, the semigroup law is proved. $\qquad\square$

The following proposition shows that condition (I.11) holds for a quite large class of semigroups and hence is indeed useful. The assertion in the Hilbert space case is based on van Casteren [45] and was used by many authors in this or some other form.

Proposition 3.10. *Let $T(\cdot)$ be a C_0-semigroup on a Banach space X. If either X is a Hilbert space or $T(\cdot)$ is an analytic semigroup, then the generator of $T(\cdot)$ satisfies condition (I.11).*

Proof. If $T(\cdot)$ is analytic, then the resolvent of its generator A satisfies $\|R(\lambda, A)\| \leq \frac{M}{|\lambda|}$ for some M and all λ in some sector $\Sigma_{c,\varphi} := \{z : \arg(z - c) \leq \varphi + \frac{\pi}{2}\}, \varphi > 0$. Therefore, $\|R^2(a + is, A)\| \leq \frac{M^2}{a^2 + s^2}$ for every $a > s(A)$ and $s \in \mathbb{R}$, and condition (I.11) follows.

Assume now that X is a Hilbert space. Then by the representation

$$R(a + is, A)x = \int_0^\infty e^{-ist} e^{-at} T(t)x \, dt$$

and Parseval's equality we have

$$\int_{-\infty}^\infty \|R(a + is, A)x\|^2 ds = \int_0^\infty e^{-2at} \|T(t)x\|^2 \, ds < \infty$$

for every $x \in X$ and every $a > \omega_0(T)$. We now take $a > s_0(A)$ and prove convergence of the above integral on the left-hand side. By the resolvent equality we have for a fixed $a_1 > \omega_0(T)$ and $M := \sup\{\|R(z, A)\| : \operatorname{Re} z \geq a\}$ that

$$\|R(a + is, A)x\| \leq \|(I + (a_1 - a)R(a + is, A))R(a_1 + is, A)x\|$$
$$\leq (1 + (a_1 - a)M)\|R(a_1 + is, A)x\|$$

and hence

$$\int_{-\infty}^{\infty} \|R(a+is, A)x\|^2 \, ds < \infty$$

by the above considerations. Note that the same holds for the adjoint semigroup $T^*(\cdot)$ and its generator A^* as well.

We now conclude that for every $a > s_0(A)$ and $x, y \in X$ by the Cauchy–Schwarz inequality

$$\int_{-\infty}^{\infty} |\langle R^2(a+is, A)x, y\rangle| \, ds \leq \int_{-\infty}^{\infty} \|R(a+is, A)x\| \cdot \|R(a-is, A^*)y\| \, ds$$

$$\leq \left(\int_{-\infty}^{\infty} \|R(a+is, A)x\|^2 ds \right)^{\frac{1}{2}} \left(\int_{-\infty}^{\infty} \|R(a+is, A^*)y\|^2 \, ds \right)^{\frac{1}{2}} < \infty.$$

So the integrability condition (I.11) is satisfied for every $a > s_0(A)$.   □

4 Positivity in $\mathcal{L}(H)$

As a preparation to Sections II.6 and III.7, we now look at the Banach algebra $\mathcal{L}(H)$, H a Hilbert space, decomposed as

$$\mathcal{L}(H) = \mathcal{L}(H)_{sa} \oplus i\mathcal{L}(H)_{sa},$$

where $\mathcal{L}(H)_{sa}$ is the real vector space of all selfadjoint operators on H. We are interested in the order structure induced by the positive semidefinite operators on $\mathcal{L}(H)_{sa}$.

4.1 Preliminaries

Recall that an operator $T \in \mathcal{L}(H)$ on a Hilbert space H is *positive semidefinite* if $\langle Tx, x \rangle \geq 0$ for every $x \in H$, in which case we write $0 \leq T$. The set $\mathcal{L}(H)_+$ of all positive semidefinite operators is a generating cone in $\mathcal{L}(H)$ and defines a vector space order by

$$S \leq T \quad \text{if} \quad 0 \leq T - S.$$

The following properties of this order will be needed later, see e.g. Pedersen [210, Proposition 3.2.9].

Lemma 4.1. (a) *If $0 \leq S \leq T$, then $\|S\| \leq \|T\|$.*

(b) *If $\{S_n\}_{n=1}^{\infty}$ is a positive, monotone increasing sequence in $\mathcal{L}(H)_{sa}$ such that*

$$\sup_{n \in \mathbb{N}} \|S_n\| < \infty,$$

then $S := \sup_{n \in \mathbb{N}} S_n$ exists and is given by $\langle Sx, x \rangle = \sup_{n \in \mathbb{N}} \langle S_n x, x \rangle = \lim_{n \to \infty} \langle S_n x, x \rangle$. In particular, S_n converges to S in the weak operator topology.

Note that the last part of (b) follows using the polarisation identity.

We now consider linear operators acting on $\mathcal{L}(H)$ respecting this order structure.

Definition 4.2. An operator $T \in \mathcal{L}(\mathcal{L}(H))$ is called *positive* (or *positivity preserving*) if $0 \leq TS$ for every $0 \leq S \in \mathcal{L}(H)$.

For a positive operator $S \in \mathcal{L}(\mathcal{L}(H))$ we again write $0 \leq S$. Analogously, we write $T \leq S$ whenever $0 \leq S - T$. Observe that such a positive operator is determined by its restriction to $\mathcal{L}(H)_{sa}$.

4.2 Spectral properties of positive operators

We now discuss some spectral and resolvent properties of positive operators on $\mathcal{L}(H)$. For the reader's convenience we present some proofs and start with the following lemma.

Lemma 4.3. *Let H be a Hilbert space and let T be a positive operator on $\mathcal{L}(H)$. If $|\mu| > r(T)$, then*

$$|\langle R(\mu, T)Sx, x \rangle| \leq \langle R(|\mu|, T)Sx, x \rangle$$

for every $x \in H$ and $0 \leq S \in \mathcal{L}(H)$.

Proof. For $x \in H$, $0 \leq S \in \mathcal{L}(H)$ and μ satisfying $|\mu| > r(T)$ we have, by the Neumann series representation for the resolvent,

$$|\langle R(\mu, T)Sx, x \rangle| \leq \sum_{k=0}^{\infty} \frac{|\langle T^k Sx, x \rangle|}{|\mu|^{k+1}} = \sum_{k=0}^{\infty} \frac{\langle T^k Sx, x \rangle}{|\mu|^{k+1}} = \langle R(|\mu|, T)Sx, x \rangle,$$

proving the assertion. $\qquad\square$

The following properties of positive operators on $\mathcal{L}(H)$ correspond to the classical Perron–Frobenius theorem for positive matrices and will be crucial for our approach to Lyapunov's equation in Sections II.6 and III.7.

Theorem 4.4. *Let T be a positive operator on $\mathcal{L}(H)$, H a Hilbert space. Then the following assertions hold.*

(a) $r(T) \in \sigma(T)$.

(b) $R(\mu, T) \geq 0$ *if and only if* $\mu > r(T)$.

In particular, $r(T) < 1$ if and only if $1 \in \rho(T)$ and $R(1, T) \geq 0$.

Proof. (a) For each $\lambda \in \sigma(T)$ such that $|\lambda| = r(T)$ we have

$$\lim_{\mu \to \lambda, \, |\mu| > r(T)} \|R(\mu, T)\| = \infty.$$

By the uniform boundedness principle there exists $S \in \mathcal{L}(H)_+$ and a sequence $\{\mu_n\}_{n=1}^{\infty}$ converging to λ such that $\lim_{n \to \infty} \|R(\mu_n, T)S\| = \infty$. This implies, by the polarisation identity, that $\limsup_{n\to\infty} |\langle R(\mu_n, T)Sx, x\rangle| = \infty$ for some $x \in H$. We obtain, by Lemma 4.3,

$$\limsup_{n\to\infty}\langle R(|\mu_n|, T)Sx, x\rangle = \infty$$

implying $|\lambda| = r(T) \in \sigma(T)$.

(b) The "if" direction follows directly from the Neumann series representation for the resolvent of T. To show the "only if" direction, assume that $R(\mu, T) \geq 0$ holds for some $\mu \in \rho(T)$. We first show that μ is real. Take some $\nu > r(T)$ with $\nu \neq \mu$. By the above "if" direction, $R(\nu, T) \geq 0$. For every $x \in H$ we obtain by the resolvent identity

$$(\mu - \nu)\langle R(\mu, T)R(\nu, T)Ix, x\rangle = \langle R(\nu, T)Ix, x\rangle - \langle R(\mu, T)Ix, x\rangle \in \mathbb{R}.$$

Since $\nu \neq \mu$ and hence $R(\nu, T)I \neq R(\mu, T)I$, we have $\langle R(\nu, T)Ix, x\rangle \neq \langle R(\mu, T)Ix, x\rangle$ for some $x \in H$. Thus the right-hand side of the above equation is real and non-zero for this x and hence $R(\mu, T)R(\nu, T) \geq 0$ implies $\mu \in \mathbb{R}$.

To show $\mu > r(T)$, we assume $\mu \leq r(T)$ and take $\nu > r(T)$. By the above, $R(\nu, T) \geq 0$ and therefore, by the resolvent identity,

$$R(\mu, T) = R(\nu, T) + (\nu - \mu)R(\mu, T)R(\nu, T) \geq R(\nu, T). \qquad (I.23)$$

By (a) we have $\lim_{\nu\to r(T)+} \|R(\nu, T)\| = \infty$ implying by the uniform boundedness principle that $\limsup_{\nu\to r(T)+} \|R(\nu, T)S\| = \infty$ for some $S \in \mathcal{L}(H)_+$. Moreover, it follows from the polarisation identity that $\limsup_{\nu\to r(T)+}\langle R(\nu, T)Sx, x\rangle = \infty$ for some $x \in H$, contradicting (I.23). $\qquad\square$

4.3   Implemented operators

We now present an important class of positive operators on $\mathcal{L}(H)$ which will play a crucial role in Section II.6.

Definition 4.5. To $T \in \mathcal{L}(H)$, H a Hilbert space, corresponds its *implemented operator* $\mathcal{T}$ on $\mathcal{L}(H)$ defined by $\mathcal{T}S := T^*ST$, $S \in \mathcal{L}(H)$.

Implemented operators have the following elementary properties.

Lemma 4.6. *For each* $T \in \mathcal{L}(H)$, *the implemented operator* $\mathcal{T}$ *is positive and satisfies* $\|\mathcal{T}\| = \|\mathcal{T}I\| = \|T\|^2$.

Proof. Positivity of $\mathcal{T}$ follows directly from its definition. For the second part observe that $\|\mathcal{T}S\| = \|T^*ST\| \leq \|T\|^2\|S\|$ for every S implies $\|\mathcal{T}\| \leq \|T\|^2$. On the other hand we have $\|\mathcal{T}\| \geq \|\mathcal{T}I\| = \|T^*T\| = \|T\|^2$, proving the assertion. $\qquad\square$

4.4 Implemented semigroups

Starting from a C_0-semigroup on a Hilbert space, we define its "implemented semigroup" and list some basic properties.

Definition 4.7. For a C_0-semigroup $T(\cdot)$ on a Hilbert space H, we call the family $(\mathcal{T}(t))_{t\geq 0}$ of operators on $\mathcal{L}(H)$ given by

$$\mathcal{T}(t)S := T^*(t)ST(t), \quad S \in \mathcal{L}(H),$$

the *implemented semigroup* corresponding to $T(\cdot)$.

The family $\mathcal{T}(\cdot)$ is indeed a semigroup, but the mapping $t \mapsto \mathcal{T}(t)Sx$ is only continuous for every $S \in \mathcal{L}(H)$ and $x \in H$, i.e., $\mathcal{T}(\cdot)$ is "strongly operator continuous". It is a C_0-semigroup on $\mathcal{L}(H)$ if and only if $T(\cdot)$ is norm continuous.

The following properties of $\mathcal{T}(\cdot)$ are well-known, see e.g. Nagel (ed.) [196, Section D-IV.2], Batty, Robinson [28, Example 2.3.7] and also Kühnemund [158].

Lemma 4.8. *For the semigroup $\mathcal{T}(\cdot)$ on $\mathcal{L}(H)$ implemented by a C_0-semigroup $T(\cdot)$ on a Hilbert space H the following assertions hold.*

(a) $\|\mathcal{T}(t)\| = \|\mathcal{T}(t)I\| = \|T(t)\|^2$.

(b) *There is a unique operator $\mathcal{A}$ with domain $D(\mathcal{A})$ in $\mathcal{L}(H)$, called the gener- ator of $\mathcal{T}(\cdot)$, with the following properties.*

    1) *$D(\mathcal{A})$ is the set of all $S \in \mathcal{L}(H)$ such that $S(D(A)) \subset D(A^*)$ and the operator $A^*S + SA : D(A) \to H$ has a continuous extension to H denoted by $\mathcal{A}S$.*

    2) *If $\int_0^\infty e^{-\lambda t}\mathcal{T}(t)Sx\,dt$ converges for every $S \in \mathcal{L}(H)$ and $x \in H$, then $\lambda \in \rho(\mathcal{A})$ and*

$$R(\lambda, \mathcal{A})Sx = \int_0^\infty e^{-\lambda t}\mathcal{T}(t)Sx\,dt.$$

In particular, this equality holds for every λ with $\operatorname{Re}\lambda > \omega_0(\mathcal{T})$.

To shorten the notation, we write $R(\lambda, \mathcal{A}) = \int_0^\infty e^{-\lambda t}\mathcal{T}(t)\,dt$ meaning that the integral exists in the above sense.

While $\mathcal{T}(\cdot)$ is not strongly continuous on $\mathcal{L}(H)$ in general, we gain posi- tivity as a new property. Indeed, $\mathcal{T}(\cdot)$ is positive on $\mathcal{L}(H)$ since $\langle \mathcal{T}(t)Sx, x\rangle = \langle ST(t)x, T(t)x\rangle \geq 0$ for every positive semidefinite operator $S \in \mathcal{L}(H)$ and every $x \in H$.

Such semigroups have been studied systematically in e.g. Nagel (ed.) [196, Section D-IV.1-2], Nagel, Rhandi [199], Groh, Neubrander [112] and Batty, Robin- son [28, Section 2]. For example, the following spectral property is analogous to Theorem 4.4 above and plays the key role in our considerations.

Theorem 4.9. *Let $(\mathcal{T}(t))_{t\geq 0}$ be an implemented semigroup on $\mathcal{L}(H)$ for a Hilbert space H with generator $\mathcal{A}$. Then $s(\mathcal{A}) \in \sigma(\mathcal{A})$, and $R(\lambda, \mathcal{A}) \geq 0$ in $\mathcal{L}(H)$ if and only if $\lambda > s(\mathcal{A})$. Moreover, $\omega_0(\mathcal{T}) = s(\mathcal{A})$ holds.*

Chapter II

Stability of linear operators

In this chapter we study power and polynomial boundedness, strong, weak and almost weak stability of linear bounded operators on Banach spaces. In many cases, these properties can be characterised through the behaviour of the resolvent of the operator in a neighbourhood of the unit circle. Similar results on the stability of C_0-semigroups will be discussed in Chapter III.

1 Power boundedness

We start by discussing power boundedness and the related property of polynomial boundedness of an operator on a Banach space. While power boundedness is fundamental for many purposes, it is difficult to check in the absence of contractivity. On the contrary, polynomial boundedness is much easier to characterise.

1.1 Preliminaries

We first introduce power bounded operators and show some elementary properties.

Definition 1.1. A linear operator T on a Banach space X is called *power bounded* if $\sup_{n\in\mathbb{N}} \|T^n\| < \infty$.

An immediate necessary spectral condition for power bounded operators is the following.

Remark 1.2. The spectral radius $r(T)$ of a power bounded operator T on a Banach space X satisfies

$$r(T) = \inf_{n\in\mathbb{N}} (\|T^n\|)^{\frac{1}{n}} \leq 1,$$

and hence $\sigma(T) \subset \{z : |z| \leq 1\}$.

Note that $r(T) \leq 1$ does not imply power boundedness as can be seen from $T = \begin{pmatrix} 1 & 1 \\ 0 & 1 \end{pmatrix}$ on $\mathbb{C}^2$. Moreover, we refer to Subsection 1.3 for more sophisticated examples and a quite complete description of the possible growth of the powers of an operator satisfying $r(T) \leq 1$, see Example 1.16 below. However, the strict inequality $r(T) < 1$ automatically implies a much stronger assertion.

Proposition 1.3. *Let X be a Banach space and $T \in \mathcal{L}(X)$. The following assertions are equivalent.*

(a) $r(T) < 1$.

(b) $\lim_{n \to \infty} \|T^n\| = 0$.

(c) *T is uniformly exponentially stable, i.e., there exist constants $M \geq 0$ and $\varepsilon > 0$ such that $\|T^n\| \leq Me^{-\varepsilon n}$ for all $n \in \mathbb{N}$.*

The proof follows from the formula $r(T) = \lim_{n \in \mathbb{N}} (\|T^n\|)^{\frac{1}{n}}$.

Remark 1.4. It is interesting that for power bounded operators on separable Banach spaces some more information on the spectrum is known. For example, Jamison [135] proved that for every power bounded operator T on a separable Banach space X the boundary point spectrum $P_\sigma(T) \cap \Gamma$ is countable. For more information on this phenomenon see Ransford [218], Ransford, Roginskaya [219] and Badea, Grivaux [13, 14].

We will see later that countability of the entire spectrum on the unit circle plays an important role for strong stability of the operator (see Subsection 2.3).

The following easy lemma is useful in order to understand power boundedness.

Lemma 1.5. *Let T be power bounded on a Banach space X. Then there exists an equivalent norm on X such that T becomes a contraction.*

Proof. Take $\|x\|_1 := \sup_{n \in \mathbb{N} \cup \{0\}} \|T^n x\|$ for every $x \in X$.   □

Remark 1.6. It is difficult to characterise those power bounded operators on Hilbert spaces which are similar to a contraction for a Hilbert space norm. Foguel [87] showed that this is not always true (see also Halmos [120]).

The next result, due to Guo, Zwart [113] and van Casteren [46], characterises power bounded operators by strong Cesàro-boundedness of the orbits.

Proposition 1.7. *An operator T on a Banach space X is power bounded if and only if*

$$\sup_{n \in \mathbb{N}} \frac{1}{n+1} \sum_{k=0}^{n} \|T^k x\|^2 < \infty \quad \text{for every } x \in X,$$

$$\sup_{n \in \mathbb{N}} \frac{1}{n+1} \sum_{k=0}^{n} \|T'^k y\|^2 < \infty \quad \text{for every } y \in X'.$$

Proof. The necessity of the above conditions is clear, so we have to prove sufficiency.

Fix $x \in X$ and $y \in X'$. By the Cauchy-Schwarz inequality we obtain

$$|\langle T^n x, y \rangle| = \frac{1}{n+1} \sum_{k=0}^{n} |\langle T^n x, y \rangle| = \frac{1}{n+1} \sum_{k=0}^{n} |\langle T^k x, T'^{(n-k)} y \rangle|$$

$$\leq \left(\frac{1}{n+1} \sum_{k=0}^{n} \|T^k x\|^2 \right)^{\frac{1}{2}} \left(\frac{1}{n+1} \sum_{k=0}^{n} \|T'^{(n-k)} y\|^2 \right)^{\frac{1}{2}}.$$

So by assumption and the uniform boundedness principle we obtain that $\sup_{n \in \mathbb{N}} \|T^n\| < \infty$. $\square$

We finish this subsection by an elementary but interesting characterisation of power bounded operators as shift operators. We first treat contractions.

Theorem 1.8. *Let T be a contraction on a Banach space X. Then T is isometrically isomorphic to the left shift on a closed subspace of $l^\infty(X)$.*

Proof. Define the operator $J : X \to l^\infty(X)$ by

$$Jx := (x, Tx, T^2 x, T^3 x, \ldots), \quad x \in X,$$

i.e., we identify x with its orbit under $(T^n)_{n=0}^{\infty}$. Since T is contractive, J is an isometry. Moreover,

$$JTx = (Tx, T^2 x, T^3 x, \ldots), \quad x \in X,$$

i.e., T corresponds to the left shift on rg J which is a closed shift-invariant subspace of $l^\infty(X)$. $\square$

Using Lemma 1.5 we now obtain an analogous characterisation of power bounded operators.

Corollary 1.9. *Let T be a power bounded operator on a Banach space X. Then T is isomorphic to the left shift on a closed subspace of $l^\infty(X_1)$, where X_1 is X endowed with an equivalent norm.*

1.2 Characterisation via resolvent

We begin with the following proposition which is analogous to one implication in the Hille–Yosida theorem for C_0-semigroups.

Proposition 1.10. *Let X be a Banach space and let $T \in \mathcal{L}(X)$ satisfy $\|T^k\| \leq M$ for some $M \geq 1$ and all $k \in \mathbb{N}_0$. Then T satisfies the strong Kreiss resolvent condition (also called iterated resolvent condition) for the constant M, i.e.,*

$$\|R^n(\lambda, T)\| \leq \frac{M}{(|\lambda| - 1)^n} \qquad \text{for all } n \in \mathbb{N} \text{ and } \lambda \text{ with } |\lambda| > 1. \qquad \text{(II.1)}$$

Proof. Observe that by the Neumann series representation $R(\lambda, T) = \sum_{k=0}^{\infty} \frac{T^k}{\lambda^{k+1}}$ for $|\lambda| > 1$, we have

$$R^n(\lambda, T) = \frac{(-1)^{n-1}}{(n-1)!} R^{(n-1)}(\lambda, T) = \frac{(-1)^{n-1}}{(n-1)!} \sum_{k=0}^{\infty} \left(\frac{1}{\lambda^{k+1}}\right)^{(n-1)} T^k$$

$$= \frac{1}{(n-1)!} \sum_{k=0}^{\infty} \frac{(k+1)\cdots(k+n-1)}{\lambda^{k+n}} T^k.$$

This implies

$$\|R^n(\lambda, T)\| \leq \frac{M}{(n-1)!} \sum_{k=0}^{\infty} \frac{(k+1)\cdots(k+n-1)}{|\lambda|^{k+n}}$$

$$= \frac{M(-1)^{n-1}}{(n-1)!} \sum_{k=0}^{\infty} \left(\frac{1}{z^{k+1}}\right)^{(n-1)} \Big|_{z=|\lambda|}$$

$$= \frac{M(-1)^{n-1}}{(n-1)!} \left(\frac{1}{z-1}\right)^{(n-1)} \Big|_{z=|\lambda|} = \frac{M}{(|\lambda|-1)^n},$$

and the proposition is proved.   □

It is surprising that the converse implication in Proposition 1.10 is not true, i.e., the discrete analogue of the Hille–Yosida theorem does not hold. For a counterexample with the maximal possible growth of $\|T^n\|$ being equal to $\sqrt{n}$, see Lubich, Nevanlinna [172]. For a systematic discussion of the strong Kreiss condition and the related uniform Kreiss condition, we refer the reader to Gomilko, Zemánek [106], Montes-Rodríguez, Sánchez-Álvarez and Zemánek [187], see also Nagy, Zemánek [200] and Nevanlinna [206].

Note that the question whether the strong Kreiss resolvent condition implies power boundedness for operators on Hilbert spaces is still open.

Condition (II.1) for $n = 1$ is called the *Kreiss resolvent condition* and plays an important role in numerical analysis. By the celebrated Kreiss matrix theorem, it is equivalent to power boundedness for operators acting on finite-dimensional spaces. On infinite-dimensional spaces this is no longer true, see Subsection 1.3 for details. We also mention here the *Ritt resolvent condition* (also called *Tadmor–Ritt resolvent condition*)

$$\|R(\lambda, T)\| \leq \frac{\text{const}}{|\lambda - 1|} \qquad \text{for all } \lambda \text{ with } |\lambda| > 1.$$

This condition does imply power boundedness, see Nagy, Zemánek [200], and also $\sigma(T) \cap \Gamma \subset \{1\}$, hence it is far from being necessary for power boundedness of general operators. We refer to Shields [232], Lubich, Nevanlinna [172], Nagy, Zemánek [200], Borovykh, Drissi, Spijker [40], Nevanlinna [206], Spijker, Tracogna,

Welfert [233], Tsedenbayar, Zemánek [241] and Vitse [246]–[248] for systematic studies of operators satisfying Kreiss and Ritt resolvent conditions.

In the following we consider conditions not involving all powers of the resolvent and characterise power boundedness at least on Hilbert spaces.

We first state an easy but very useful lemma.

Lemma 1.11. *Let X be a Banach space and $T \in \mathcal{L}(X)$. Then*

$$T^n = \frac{r^{n+1}}{2\pi} \int_0^{2\pi} e^{i\varphi(n+1)} R(re^{i\varphi}, T)d\varphi = \frac{r^{n+2}}{2\pi(n+1)} \int_0^{2\pi} e^{i\varphi(n+1)} R^2(re^{i\varphi}, T)d\varphi$$

(II.2)

for every $n \in \mathbb{N}$ and $r > r(T)$.

Proof. The first equality in (II.2) follows directly from the Dunford functional calculus and the second by integration by parts. □

The main result of this subsection is the following theorem which is a discrete analogue of a characterisation of bounded C_0-semigroups due to Gomilko [104] and Shi, Feng [231], see Theorem III.1.11.

Theorem 1.12. *Let X be a Banach space and $T \in \mathcal{L}(X)$ with $r(T) \leq 1$. Consider the following assertions.*

(a) $\limsup_{r\to 1+}(r-1) \int_0^{2\pi} \|R(re^{i\varphi}, T)x\|^2 d\varphi < \infty$ for all $x \in X$,
   $\limsup_{r\to 1+}(r-1) \int_0^{2\pi} \|R(re^{i\varphi}, T')y\|^2 d\varphi < \infty$ for all $y \in X'$;

(b) $\limsup_{r\to 1+}(r-1) \int_0^{2\pi} |\langle R^2(re^{i\varphi}, T)x, y\rangle| d\varphi < \infty$ for all $x \in X$, $y \in X'$;

(c) T *is power bounded.*

Then (a)$\Rightarrow$(b)$\Rightarrow$(c). *Moreover, if X is a Hilbert space, then* (c)$\Rightarrow$(a), *hence* (a), (b) *and* (c) *are equivalent.*

Proof. By the Cauchy-Schwarz inequality we have

$$\int_0^{2\pi} |\langle R^2(re^{i\varphi}, T)x, y\rangle| d\varphi = \int_0^{2\pi} |\langle R(re^{i\varphi}, T)x, R(re^{i\varphi}, T')y\rangle| d\varphi$$

$$\leq \left(\int_0^{2\pi} \|R(re^{i\varphi}, T)x\|^2 d\varphi \right)^{\frac{1}{2}} \left(\int_0^{2\pi} \|R(re^{i\varphi}, T')y\|^2 d\varphi \right)^{\frac{1}{2}}$$

for all $x \in X$, $y \in X'$ and $r > r(T)$. This proves the implication (a)$\Rightarrow$(b).

For the implication (b)$\Rightarrow$(c) take $n \in \mathbb{N}$ and $r > 1$. By Lemma 1.11 we have

$$|\langle T^n x, y\rangle| \leq \frac{r^{n+2}}{2\pi(n+1)} \int_0^{2\pi} |\langle R^2(re^{i\varphi})x, y\rangle| d\varphi$$

for every $x \in X$ and $y \in X'$. By (b) and the uniform boundedness principle there exists a constant $M > 0$ such that

$$(r-1) \int_0^{2\pi} |\langle R^2(re^{i\varphi}, T)x, y\rangle| d\varphi \leq M\|x\|\|y\| \quad \text{for every } x \in X, \ y \in X' \text{ and } r > 1.$$

Therefore we obtain

$$|\langle T^n x, y\rangle| \le \frac{M r^{n+2}}{2\pi(n+1)(r-1)}\|x\|\|y\|. \tag{II.3}$$

Take now $r := 1 + \frac{1}{n+1}$. Then $\frac{r^{n+2}}{(n+1)(r-1)} = \left(1 + \frac{1}{n+1}\right)^{n+2} \to e$ as $n \to \infty$, and we obtain by (II.3) that $\sup_{n \in \mathbb{N}} \|T^n\|$ is finite.

For the second part of the theorem assume that X is a Hilbert space and T is power bounded. Then, by Lemma 1.11 and Parseval's equality,

$$(r-1)\int_0^{2\pi} \|R(re^{i\varphi}, T)x\|^2 d\varphi = \frac{r-1}{r^2}\sum_{n=0}^{\infty}\frac{\|T^n x\|^2}{r^{2n}} = \frac{1}{r+1}(1-s)\sum_{n=0}^{\infty} s^n \|T^n x\|^2 \tag{II.4}$$

for $s := \frac{1}{r^2} < 1$. Note that the right-hand side of (II.4) is, up to the factor $1/(r+1)$, the Abel mean of the sequence $\{\|T^n x\|^2\}_{n=0}^{\infty}$, so it is bounded because of the power boundedness of T. This proves the first part of (a).

Analogously, we obtain the second part of (a) using the power boundedness of T'. $\qquad\square$

Remarks 1.13. 1) As can be seen from the proof, Theorem 1.12 can also be formulated for a single weak orbit $\{\langle T^n x, y\rangle : n \in \mathbb{N}\}$. More precisely, for a fixed pair $x \in X$ and $y \in X'$, condition (a) implies (b), (b) implies boundedness of the corresponding weak orbit and the converse implications hold for Hilbert spaces.

2) Moreover, one can replace condition (a) by

$$\limsup_{r\to 1+}(r-1)\int_0^{2\pi} \|R(re^{i\varphi}, T)x\|^p d\varphi < \infty,$$

$$\limsup_{r\to 1+}(r-1)\int_0^{2\pi} \|R(re^{i\varphi}, T')y\|^q d\varphi < \infty$$

for some $p, q > 1$ (depending on x and y) with $\frac{1}{p} + \frac{1}{q} = 1$.

We now show that in Theorem 1.12 condition (c) does not imply (a) in general, not even for isometric invertible operators.

Example 1.14. Take $X = l^\infty$ and T given by

$$T(x_1, x_2, x_3, \dots) := (a x_1, a^2 x_2, a^3 x_3, \dots)$$

for some $a \in \Gamma$ with angle rationally independent of π. The operator T is isometric and hence contractive, invertible and satisfies $\sigma(T) = \Gamma$. Moreover, its resolvent is given by

$$R(\lambda, T)(x_1, x_2, x_3, \dots) = \left(\frac{x_1}{\lambda - a}, \frac{x_2}{\lambda - a^2}, \frac{x_3}{\lambda - a^3}, \dots\right), \quad |\lambda| \ne 1.$$

So we have

$$\|R(\lambda, T)\mathbf{1}\| = \sup_{n \in \mathbb{N}} \frac{1}{|\lambda - a^n|} = \frac{1}{\operatorname{dist}(\lambda, \Gamma)} = \frac{1}{|\lambda| - 1}, \quad |\lambda| > 1,$$

which implies

$$(r-1)\int_0^{2\pi} \|R(re^{i\varphi}, T)\mathbf{1}\|^2 d\varphi = (r-1)\int_0^{2\pi} \frac{1}{(r-1)^2} d\varphi = \frac{2\pi}{r-1} \to \infty \quad \text{as } r \to 1+.$$

This shows that condition (a) in Theorem 1.12 does not hold for T. In view of Remark 1.13.2) we note also that $\lim_{r \to 1+}(r-1)\int_0^{2\pi} \|R(re^{i\varphi}, T)x\|^p d\varphi = \infty$ for every $1 < p < \infty$.

The pre-adjoint of T on l^1 also does not satisfy (a). We remark that this operator is again isometric and invertible.

A useful characterisation of power boundedness on Banach spaces is still unknown.

1.3 Polynomial boundedness

In this subsection we discuss the related notion of polynomial boundedness which, surprisingly, is much easier to characterise.

Definition 1.15. A bounded linear operator T on a Banach space X is called *polynomially bounded* if $\|T^n\| \le p(n)$ for some polynomial p and all $n \in \mathbb{N}$.

Without loss of generality we will assume the polynomial to be of the form $p(t) = Ct^d$.

Note that a polynomially bounded operator T again satisfies $r(T) \le 1$. The following example shows that the converse implication is far from being true.

Example 1.16 (Operators satisfying $r(T) \le 1$ with non-polynomial growth). Consider the Hilbert space

$$H := l_a^2 = \left\{ \{x_n\}_{n=1}^\infty \subset \mathbb{C} : \sum_{n=1}^\infty |x_n|^2 a_n^2 < \infty \right\}$$

for a positive sequence $\{a_n\}_{n=1}^\infty$ satisfying

$$a_{n+m} \le a_n a_m \quad \text{for all } n, m \in \mathbb{N} \tag{II.5}$$

and with the natural scalar product. On H take the right shift operator

$$T(x_1, x_2, x_3, \ldots) := (0, x_1, x_2, \ldots).$$

Then for $x = (x_1, x_2, \ldots) \in H$ we have by (II.5)

$$\|T^k x\|^2 = \|(0, \ldots, 0, x_1, x_2, \ldots)\|^2 = \sum_{n=1}^\infty a_{n+k}^2 |x_n|^2 \le a_k^2 \sum_{n=1}^\infty a_n^2 |x_n|^2 = a_k^2 \|x\|^2$$

for every $k \in \mathbb{N}$. Moreover, $\|T^k e_1\| = \|e_{k+1}\| = a_{k+1} = \frac{a_{k+1}}{a_1}\|e_1\|$, where e_k denotes the sequence having the k-th component equal to 1 and all others equal to zero. Therefore we have the norm estimate

$$\frac{a_{k+1}}{a_1} \le \|T^k\| \le a_k,$$

which implies that the powers of T have the same growth as the sequence $\{a_n\}_{n=1}^{\infty}$. Now every sequence satisfying (II.5) and growing faster than every polynomial but slower than any exponential function with positive exponent yields an operator growing non-polynomially but with $r(T) \le 1$.

As a concrete example of such a sequence take

$$a_n := (n+5)^{\ln(n+5)} = e^{\ln^2(n+5)}.$$

The assertion about the growth is clear and we only need to check condition (II.5). For $n, m \ge 6$ we have to prove that $\ln^2(n+m) \le \ln^2 n + \ln^2 m$. This follows from the following two properties of the function $x \mapsto \ln^2 x$:

$$f(2x) \le 2f(x), \tag{II.6}$$
$$f''(x) < 0 \tag{II.7}$$

satisfied for $x \ge 6$. Indeed, from the conditions above we see that the inequality

$$f(x+y) - f(x) \le f(y)$$

holds for $x = y$. For a fixed y the derivative of the left-hand side is negative, so the inequality holds for all $x, y \ge 6$. To finish, we mention that, for the function $f : x \mapsto \ln^2 x$, condition (II.7) is immediate and condition (II.6) follows from the fact that the inequality $\ln^2(2x) \le 2\ln^2(x)$ is equivalent to $2x \le x^{\sqrt{2}}$ which is satisfied for $x \ge 6$. So we have constructed an operator whose powers grow as $n^{\ln n}$.

Analogously one can construct an operator with powers growing as $n^{(\ln n)^{\alpha}}$ for any $\alpha \ge 1$.

The following characterisation of polynomial boundedness uses the resolvent of T in a neighbourhood of the unit circle, see Eisner, Zwart [73]. See also Lubich, Nevanlinna [172] for a related result involving all powers of the resolvent.

Theorem 1.17. *Let T be a bounded operator on a Banach space X with $r(T) \le 1$.*
If

$$\limsup_{|z|\to 1+}(|z|-1)^d\|R(z,T)\| < \infty \quad \text{for some } d \ge 0, \tag{II.8}$$

then

$$\|T^n\| \le Cn^d \quad \text{for some } C > 0 \text{ and all } n \in \mathbb{N}. \tag{II.9}$$

Moreover, if (II.9) holds for $d = k$, then (II.8) holds with $d = k+1$.

Proof. Assume that condition (II.8) holds and take $n \in \mathbb{N}$ and $r > 1$. By Lemma 1.11 and (II.8) we have

$$\|T^n\| \leq \frac{r^{n+1}}{2\pi} \int_0^{2\pi} \|R(re^{i\varphi}, T)\| d\varphi \leq \frac{Mr^{n+1}}{(r-1)^d}$$

for $M > \limsup_{|z| \to 1+} (|z|-1)^d \|R(z,T)\|$ and r close enough to 1. Taking $r := 1+\frac{1}{n}$ for n large enough we obtain $\|T^n\| \leq 2Men^d$ and the first part of the theorem is proved.

For the second part we assume that condition (II.9) holds for $d = k$. Take $n \in \mathbb{N}$, $r > 1$, $\varphi \in [0, 2\pi)$, and $q := \frac{1}{r} < 1$. Then

$$\|R(re^{i\varphi}, T)\| \leq \sum_{n=0}^{\infty} \frac{\|T^n\|}{r^{n+1}} \leq Cq \sum_{n=0}^{\infty} n^k q^n \leq C\sum_{n=0}^{k-1} n^k + C\sum_{n=k}^{\infty} n^k q^n$$

$$\leq C\sum_{n=0}^{k-1} n^k + C\tilde{C}q^k \frac{d^k}{dq^k} \sum_{n=0}^{\infty} q^n \leq C\sum_{n=0}^{k-1} n^k + \frac{C\tilde{C}k!}{(1-q)^{k+1}},$$

where $\tilde{C}$ is such that $n^k \leq \tilde{C} \cdot n(n-1) \cdot \ldots \cdot (n-k+1)$ for all $n > k$. For $k = 0$ we suppose the first sum on the right-hand side to be equal to zero. Substituting q by $\frac{1}{r}$ we obtain condition (II.8) for $d = k+1$. $\qquad\square$

Remark 1.18. Note that, by the inequality $\text{dist}(\lambda, \sigma(T)) \geq \frac{1}{\|R(\lambda,T)\|}$, condition (II.8) for $0 \leq d < 1$ already implies $r(T) < 1$ and hence uniform exponential stability. So for $0 \leq d < 1$ Theorem 1.17 does not give the best information about the growth of the powers. Nevertheless, for $d = 1$, i.e., for the above mentioned Kreiss resolvent condition, the growth stated in Theorem 1.17 is the best possible and the exponent d in (II.9) cannot be decreased in general, see Shields [232]. For $d > 1$ it is not clear whether Theorem 1.17 is optimal.

2   Strong stability

In this section we consider a weaker stability concept than the norm stability discussed in Proposition 1.3 and replace uniform by pointwise convergence.

2.1   Preliminaries

Let us introduce strongly stable operators and present some of their fundamental properties.

Definition 2.1. An operator T on a Banach space X is called *strongly stable* if $\lim_{n \to \infty} \|T^n x\| = 0$ for every $x \in X$.

The first part of the following example is, in a certain sense, typical for Hilbert spaces (see Theorem 2.11 below).

Example 2.2. (a) Consider $H := l^2(\mathbb{N}, H_0)$ for a Hilbert space H_0 and $T \in \mathcal{L}(H)$ defined by

$$T(x_1, x_2, x_3, \ldots) := (x_2, x_3, \ldots). \tag{II.10}$$

The operator T, called the *left shift* on H, is strongly stable.

Analogously, the operator defined by formula (II.10) is also strongly stable on the spaces $c_0(\mathbb{N}, X)$, $l^p(\mathbb{N}, X)$ for a Banach space X and $1 \le p < \infty$, but not on $l^\infty(\mathbb{N}, X)$.

(b) Consider $X := l^p$ for $1 \le p < \infty$ and the multiplication operator $T \in \mathcal{L}(X)$ defined by

$$T(x_1, x_2, x_3, \ldots) := (a_1 x_1, a_2 x_2, a_3 x_3, \ldots)$$

for a sequence $(a_n)_{n=1}^\infty \subset \{z : |z| < 1\}$. By the density of vectors with finitely many non-zero coordinates, the operator T is strongly stable. Note that by Proposition 1.3, T fails to be uniformly exponentially stable if $\sup_n |a_n| = 1$.

Analogously, such a multiplication operator is also strongly stable on c_0 but not on l^∞.

The following property of strongly stable operators is an easy consequence of the uniform boundedness principle.

Remark 2.3. Every strongly stable operator T on a Banach space X is power bounded which in particular implies $\sigma(T) \subset \{z : |z| \le 1\}$. Moreover, the properties $P_\sigma(T) \cap \Gamma = \emptyset$ and $P_\sigma(T') \cap \Gamma = \emptyset$ are necessary for strong stability.

We now present an elementary property which is very helpful to show strong stability.

Lemma 2.4. Let X be a Banach space, $T \in \mathcal{L}(X)$ power bounded and $x \in X$.

(a) If there exists a subsequence $\{n_k\}_{k=1}^\infty \subset \mathbb{N}$ such that $\lim_{k \to \infty} \|T^{n_k} x\| = 0$, then $\lim_{n \to \infty} \|T^n x\| = 0$.

(b) If T is a contraction, then there exists $\lim_{n \to \infty} \|T^n x\|$.

Proof. The second assertion follows from the fact that for a contraction the sequence $\{\|T^n x\|\}_{n=1}^\infty$ is non-increasing. For the first one take $\varepsilon > 0$, $M := \sup_{n \in \mathbb{N}} \|T^n\|$ and $k \in \mathbb{N}$ such that $\|T^{n_k} x\| \le \varepsilon$. Then we have

$$\|T^n x\| \le \|T^{n-n_k}\| \|T^{n_k} x\| \le M\varepsilon$$

for every $n \ge n_k$, and (a) is proved. $\square$

Remark 2.5. Assertion (b) in the above lemma is no longer true for power bounded operators. Here is an example. Consider the Hilbert space of all l^2-sequences endowed with the norm

$$\|x\| := \left(\sum_{n=1}^\infty \left[|x_{2n-1}|^2 + \frac{1}{4} |x_{2n}|^2 \right] \right)^{\frac{1}{2}}.$$

On this space consider the right shift operator which is clearly power bounded. We see that for the vector $e_1 = (1, 0, 0, \ldots)$ we have $\|T^{2n-1}e_1\| = \frac{1}{2}$ and $\|T^{2n}e_1\| = 1$ for every $n \in \mathbb{N}$ which implies

$$\frac{1}{2} = \liminf_{n \to \infty} \|T^n e_1\| \neq \limsup_{n \to \infty} \|T^n e_1\| = 1.$$

The following is a direct consequence of Lemma 2.4.

Corollary 2.6. *A power bounded operator T on a Banach space X is strongly stable if and only if*

$$\lim_{n \to \infty} \frac{1}{n+1} \sum_{k=0}^{n} \|T^k x\|^2 = 0$$

for every x in a dense subset of X.

One can also characterise strong stability without using power boundedness. This is the following result due to Zwart [263].

Proposition 2.7. *An operator T on a Banach space X is strongly stable if and only if*

$$\lim_{n \to \infty} \frac{1}{n+1} \sum_{k=0}^{n} \|T^k x\|^2 = 0 \quad \text{for all } x \in X \text{ and} \tag{II.11}$$

$$\sup_{n \in \mathbb{N}} \frac{1}{n+1} \sum_{k=0}^{n} \|T'^k y\|^2 < \infty \quad \text{for all } y \in X'. \tag{II.12}$$

The proof is analogous to the one of Proposition 1.7.

Note that one can formulate Corollary 2.6 and Proposition 2.7 for single orbits. More precisely, $\lim_{n \to \infty} T^n x = 0$ if and only if conditions (II.11) and (II.12) hold for every $y \in X'$. In particular, for a power bounded operator and $x \in X$, $\lim_{n \to \infty} T^n x = 0$ is equivalent to (II.11).

Finally, we state a surprising and deep result of Müller [188] on the asymptotic behaviour of operators which are not uniformly exponentially stable.

Theorem 2.8 (V. Müller). *Let T be a bounded linear operator on a Banach space X with $r(T) \geq 1$. For every $\{\alpha_n\}_{n=1}^{\infty} \subset [0, 1)$ converging monotonically to 0 there exists $x \in X$ with $\|x\| = 1$ such that*

$$\|T^n x\| \geq \alpha_n \quad \text{for every } n \in \mathbb{N}.$$

In other words, the orbits of a strongly but not uniformly exponentially stable operator decrease arbitrarily slowly. We refer to Müller [188], [189] and [191, Section V.37] for more details and results.

Up to now, there is no general characterisation of strong stability for operators on Banach spaces, while on Hilbert spaces there is a resolvent condition (see Subsection 2.4).

2.2   Representation as shift operators

We now present a characterisation analogous to Theorem 1.8 of strongly stable operators as shifts.

Proposition 2.9. *Let T be a strongly stable contraction on a Banach space X. Then T is isometrically isomorphic to the left shift on a closed subspace of $c_0(X)$.*

This result follows directly from Theorem 1.8. Moreover, by the renorming procedure we again obtain a characterisation for non-contractive strongly stable operators.

Corollary 2.10. *A strongly stable operator T on a Banach space X is isomorphic to the left shift on a closed shift-invariant subspace of $c_0(X_1)$, where X_1 is X endowed with an equivalent norm.*

For contractions on Hilbert spaces we have the following classical result of Foiaş [85] and de Branges–Rovnyak [43] (see also Sz.-Nagy, Foiaş [238, p. 95]).

Theorem 2.11. *Let T be a strongly stable contraction on a Hilbert space H. Then T is unitarily isomorphic to a left shift, i.e., there is a Hilbert space H_0 and a unitary operator $U : H \to H_1$ for some closed subspace $H_1 \subset l^2(\mathbb{N}, H_0)$ such that UTU^{-1} is the left shift on H_1.*

Proof. As in Proposition 2.9, the idea is to identify a vector x with the sequence $\{T^n x\}_{n=0}^{\infty}$ in an appropriate sequence space.

Observe first that by strong stability we have the equality

$$\|x\|^2 = \sum_{n=0}^{\infty} (\|T^n x\|^2 - \|T^{n+1} x\|^2). \tag{II.13}$$

We now define on H a new seminorm by

$$\|x\|_Y^2 := \|x\|^2 - \|Tx\|^2, \quad x \in H,$$

corresponding to the scalar semiproduct $\langle x, y \rangle_Y := \langle x, y \rangle - \langle Tx, Ty \rangle$. Note that contractivity of T implies the non-negativity of $\|\cdot\|_Y$. Let $H_0 := \{x : \|x\|_Y = 0\}$ and finally take the completion $Y := (H/H_0, \|\cdot\|_Y)\tilde{\ }$.

Define now the operator $J : H \to l^2(Y)$ by

$$x \mapsto (T^n x)_{n=0}^{\infty},$$

which identifies x with its orbit under $(T^n)_{n=0}^{\infty}$. By (II.13), J is an isometry and hence a unitary operator from H to its (closed) range. Observe finally that the operator JTJ^{-1} acts as the left shift on rg J. $\square$

2.3 Spectral conditions

In this subsection we discuss sufficient conditions for strong stability of an operator in terms of its spectrum. All results presented here are based on some "smallness" of the part of the spectrum on Γ.

The origin of this spectral approach to stability is the following classical "$T = I$" theorem of Gelfand [95]. The proof we give here is due to Allan, Ransford [6].

Theorem 2.12 (Gelfand, 1941). *Let X be a Banach space and $T \in \mathcal{L}(X)$. If $\sigma(T) = \{1\}$ and $\sup_{n \in \mathbb{Z}} \|T^n\| < \infty$, then $T = I$.*

Proof. Since (the principle branch of) the logarithm is holomorphic in a neighbourhood of $\sigma(T) = \{1\}$, we can define $S := -i \log T$ using Dunford's functional calculus. This operator satisfies $T = e^{iS}$ and $\sigma(S) = 0$ by the spectral mapping theorem. We show that $S = 0$.

Take $n \in \mathbb{N}$ and consider $\sin(nS) := \frac{1}{2i}(e^{inS} - e^{-inS}) = \frac{1}{2i}(T^n - T^{-n})$. We observe that this operator is power bounded satisfying

$$\|[\sin(nS)]^k\| = \left\|\left(\frac{T^n - T^{-n}}{2i}\right)^k\right\| \leq \sup_{n \in \mathbb{Z}} \|T^n\| \quad \text{for every } k \in \mathbb{N}.$$

Consider now the Taylor series representation $\sum_{k=0}^{\infty} c_k z^k$ of the principal branch of arcsin in 0 (recall that $\sigma(nS) = \{0\}$). Since $c_k \geq 0$ and $\sum_{k=0}^{\infty} c_k = \arcsin(1) = \frac{\pi}{2}$, we obtain by the above that

$$\|nS\| = \|\arcsin(\sin(mS))\| \leq \sum_{k=0}^{\infty} c_k \left\|(\sin(nS))^k\right\| \leq \frac{\pi}{2} \sup_{n \in \mathbb{Z}} \|T^n\| \quad \text{for every } n \in \mathbb{N}.$$

This implies $S = 0$ and $T = e^{iS} = I$. $\qquad\qquad\square$

As a second tool, we introduce the so-called isometric limit operator, an elegant construction due to Lyubich, Vũ [180].

For a contraction T on a Banach space X define

$$Y := \{x \in X : \lim_{n \to \infty} \|T^n x\| = 0\}.$$

Observe that Y is a closed T-invariant subspace. Consider the space $(X/Y, \|\cdot\|_T)$ for the norm $\|x + Y\|_T := \lim_{n \to \infty} \|T^n x\|$ which is well defined by Lemma 2.4. Note that $\|x + Y\|_T \leq \|x\|$ holds for every $x \in X$ by contractivity of T. Define the operator S on X/Y by

$$S(x + Y) := Tx + Y.$$

Consider finally the completion $Z := (X/Y, \|\cdot\|_T)^\sim$.

Remark 2.13. The space $(X/Y, \|\cdot\|_T)$ is not complete in general. This can be seen taking the space $X = \{(x_k)_{k=-\infty}^\infty \in l^1(\mathbb{Z}) : \sum_{k=1}^\infty k|x_k| < \infty\}$ with norm given by $\|(x_k)_{-\infty}^\infty\| := \sum_{k=-\infty}^0 |x_k| + \sum_{k=1}^\infty k|x_k|$ and the left shift operator T. Then T is contractive and satisfies

$$\|x\|_T = \lim_{n\to\infty} \|T^n x\| = \|x\|_1 \quad \text{for every } x \in X.$$

Therefore we have $(X/Y, \|\cdot\|_T) = (X, \|\cdot\|_1)$ and its completion is $Z = l^1(\mathbb{Z})$.

We now need the following lemma, see Lyubich, Vũ [180] or Engel, Nagel [78, pp. 263–264] in the continuous case.

Lemma 2.14. *Let T be a contraction on a Banach space X and take S on X/Y as above. Then S extends to an isometry on Z, called the isometric limit operator, and satisfies $\sigma(S) \subset \sigma(T)$.*

Proof. Observe first that $\|S(x+Y)\|_T = \lim_{n\to\infty} \|T^{n+1}x\| = \|x\|_T$ and hence S extends to an isometry on Z.

Take now $\lambda \in \rho(T)$ and define operators $R(\lambda)$ on $(X/Y, \|\cdot\|_T)$ by

$$R(\lambda)(x+Y) := R(\lambda, T)x + Y, \quad x \in X.$$

Since $\|x+Y\|_T \le \|x\|$ for every $x \in X$, $R(\lambda)$ extends to a bounded operator on Z again denoted by $R(\lambda)$. By definition we have $(\lambda - S)R(\lambda) = I$ and $R(\lambda)(\lambda - S) = I$ on X/Y and hence $\lambda \in \rho(S)$ and $R(\lambda) = (\lambda - S)^{-1}$. $\qquad\square$

The following theorem is the starting point for many spectral characterisations of strong stability, see Katznelson, Tzafriri [147]. Our proof of the non-trivial implication is due to Vũ [249].

Theorem 2.15 (Katznelson–Tzafriri, 1986). *Let T be a power bounded operator on a Banach space X. Then $\|T^{n+1} - T^n\| \to 0$ as $n \to \infty$ if and only if $\sigma(T) \cap \Gamma \subset \{1\}$.*

Proof. Assume first that there exists $1 \ne \lambda \in \Gamma$ with $\lambda \in \sigma(T)$. Then we have, by the spectral mapping theorem for polynomials,

$$\|T^{n+1} - T^n\| \ge r(T^{n+1} - T^n) \ge |\lambda^{n+1} - \lambda^n| = |\lambda - 1|$$

and hence $\liminf_{n\to\infty} \|T^{n+1} - T^n\| > 0$.

To show the converse implication, we first show that the condition $\sigma(T) \cap \Gamma \subset \{1\}$ and the power boundedness of T imply

$$\lim_{n\to\infty} \|T^{n+1}x - T^n x\| = 0 \quad \text{for every } x \in X. \tag{II.14}$$

By Lemma 1.5 we may assume that T is a contraction. Let S be its isometric limit operator. Then S is an isometry with $\sigma(S) \cap \Gamma \subset \{1\}$ by Lemma 2.14. Since an isometry is non-invertible if and only if its spectrum is the entire unit disc, see, e.g., Conway [52, Exercise VII.6.7], S must be invertible with S^{-1} being an isometry as well. This shows that $\sigma(S) = \{1\}$, and by Gelfand's Theorem 2.12, $S = I$ on Z. By the definition of S, we have that $x - Tx \in Y = \{y \in X : \lim_{n\to\infty} \|T^n y\| = 0\}$, and (II.14) is proved.

It remains to show that $\lim_{n\to\infty} \|T^{n+1} - T^n\| = 0$. Consider the operator U on $\mathcal{L}(X)$ given by $UR = TR$ for all $R \in \mathcal{L}(X)$. This operator is power bounded and satisfies $\sigma(U) \subset \sigma(T)$. By the above considerations we have $\lim_{n\to\infty} \|U^{n+1}R - U^n R\| = 0$ for all $R \in \mathcal{L}(X)$. It suffices to take $R := I$. $\qquad\square$

For alternative proofs of the Katznelson–Tzafriri theorem using complex analysis see Allan, Ransford [6] and Allan, Farrell, Ransford [7], as well as the original proof of Katznelson, Tzafriri [147] using harmonic analysis.

Remark 2.16. There are various generalisations and extensions of the Katznelson–Tzafriri theorem in which $T^{n+1} - T^n$ is replaced by $T^n f(T)$ for some function f. For example, Esterle, Strouse, Zouakia [80] showed that for a contraction T on a Hilbert space and for $f \in \mathcal{H}(\mathbb{D}) \cap C(\Gamma)$, the condition $\lim_{n\to\infty} \|T^n f(T)\| = 0$ holds if and only if $f = 0$ on $\sigma(T) \cap \Gamma$. This can be generalised to "polynomially bounded" operators (in the sense of von Neumann's inequality, see, e.g., Pisier [213, 214] and Davidson [56]) and to holomorphic functions on $\mathbb{D}$ and completely non-unitary contractions, see Kérchy, van Neerven [148] and Bercovici [30], respectively. The version of the Katznelson–Tzafriri theorem for C_0-semigroups is in Esterle, Strouse, Zouakia [81] and Vũ [250]. For further generalisations and history we refer to Chill, Tomilov [50, Section 5].

An immediate corollary concerning strong stability is the following.

Corollary 2.17. *Let T be a power bounded operator on a* <u>Banach</u> *space X with $\sigma(T) \cap \Gamma \subset \{1\}$. Then $\|T^n x\| \to 0$ as $n \to \infty$ for every $x \in$ <u>rg$(I - T)$</u>. In particular, $1 \notin P_\sigma(T')$ implies that T is strongly stable.*

On the basis of the Katznelson–Tzafriri theorem, Arendt, Batty [9] and independently Lyubich, Vũ [180] proved a sufficient condition for strong stability assuming countability of $\sigma(T)$ on the unit circle. The following proof using the isometric limit operator is due to Lyubich and Vũ [180], while Arendt, Batty [9] used Laplace transform techniques.

Theorem 2.18 (Arendt–Batty–Lyubich–Vũ, 1988). *Let T be power bounded on a Banach space X. Assume that*

(i) $P_\sigma(T') \cap \Gamma = \emptyset$;

(ii) $\sigma(T) \cap \Gamma$ *is countable.*

Then T is strongly stable.

Proof. We first note that, by Lemma 1.5, we can assume T to be contractive.

Assume that T is not strongly stable. Then the space $Z = (X/Y, \|\cdot\|_T)^\sim$ constructed above is non-zero. Consider now the isometric limit operator S on Z. By (ii) and Lemma 2.14, $\sigma(S) \cap \Gamma$ is still countable implying $\sigma(S) \cap \Gamma \neq \Gamma$. Since the spectrum of a noninvertible isometry is the whole unit disc, S is invertible and $\sigma(S) \subset \Gamma$ holds. So $\sigma(S)$ is a countable complete metric space and therefore contains an isolated point $\lambda_0 \in \Gamma$ by Baire's theorem.

We now show that λ_0 is an eigenvalue of S. Consider the spectral projection P of S corresponding to λ_0 and the restriction S_0 of S on $\operatorname{rg} P =: Z_0$. Since $\sigma(S_0) = \{\lambda_0\}$ and $\|S_0^n\| = 1$ for every $n \in \mathbb{Z}$, it follows from Gelfand's theorem (Theorem 2.12) that $S_0 = \lambda_0 I$ and therefore in particular $\lambda_0 \in P_\sigma(S)$.

It remains to show that $\lambda_0 \in P_\sigma(T')$ to obtain a contradiction to (i). Take $z' \in Z_0'$ and consider $0 \neq x' \in X'$ defined by

$$\langle x, x' \rangle := \langle P(x + Y), z' \rangle, \quad x \in X.$$

Then

$$\begin{aligned}
\langle x, T'x' \rangle = \langle Tx, x' \rangle &= \langle P(Tx + Y), z' \rangle \\
&= \langle PS(x + Y), z' \rangle = \lambda_0 \langle P(x + Y), z' \rangle = \lambda_0 \langle x, x' \rangle
\end{aligned}$$

for every $x \in X$ and hence $T'x' = \lambda_0 x'$ contradicting (i). $\qquad \square$

Remark 2.19. For operators with relatively weakly compact orbits (in particular, for power bounded operators on reflexive Banach spaces) condition (i) is equivalent to $P_\sigma(T) \cap \Gamma = \emptyset$.

As one of many applications of the above theorem we present the following stability result for positive operators.

Corollary 2.20. Let T be a positive power bounded operator on a Banach lattice. Then T is strongly stable if $P_\sigma(T') \cap \Gamma = \emptyset$ and $\sigma(T) \cap \Gamma \neq \Gamma$.

The proof follows from Theorem 2.18 and the Perron–Frobenius theory stating that in this case the boundary spectrum $\sigma(T) \cap \Gamma$ is multiplicatively cyclic (see Schaefer [227, Section V.4]), and hence, since different from Γ, must be a finite union of roots of unity.

The following result is an extension of the Arendt–Batty–Lyubich–Vũ theorem for completely non-unitary contractions on Hilbert spaces (for the definition of completely non-unitary operators see Remark 3.10) below.

Theorem 2.21 (Foiaş, Sz.-Nagy [238, Prop. II.6.7]). Let T be a completely non-unitary contraction on a Hilbert space H. If

$$\sigma(T) \cap \Gamma \text{ has Lebesgue measure } 0,$$

then T and T^* are both strongly stable.

See also Kérchy, van Neerven [148] for related results.

In the above theorem it does not suffice to assume contractivity and emptiness of the point spectrum. In Example III.3.19 we present a unitary and hence not strongly stable operator T satisfying $\lambda(\sigma(T)) = 0$ for the Lebesgue measure λ and $P_\sigma(T) = \emptyset$.

Remark 2.22. Although the results above have many useful applications, the smallness conditions on the boundary spectrum are far from being necessary for strong stability. Indeed, we saw in Example 2.2(b) that for a strongly stable operator the boundary spectrum can be an arbitrary closed subset of Γ. However, on superreflexive Banach spaces, countability of the boundary spectrum is equivalent to a stronger property called *superstability*, see Nagel, Räbiger [198].

2.4 Characterisation via resolvent

In this subsection we pursue a resolvent approach to stability introduced by Tomilov [243].

Our main result is the following discrete analogue to the spectral characterisation given by Tomilov. For related results we refer to his paper [243].

Theorem 2.23. *Let X be a Banach space and $T \in \mathcal{L}(X)$ with $r(T) \leq 1$ and $x \in X$. Consider the following assertions.*

(a) $\lim\limits_{r \to 1+} (r - 1) \displaystyle\int_0^{2\pi} \|R(re^{i\varphi}, T)x\|^2 \, d\varphi = 0$ *and*

$\limsup\limits_{r \to 1+} (r - 1) \displaystyle\int_0^{2\pi} \|R(re^{i\varphi}, T')y\|^2 \, d\varphi < \infty$ *for all $y \in X'$.*

(b) $\lim\limits_{n \to \infty} \|T^n x\| = 0.$

Then (a) *implies* (b). *Moreover, if X is a Hilbert space, then* (a)$\Leftrightarrow$(b).

In particular, condition (a) *for all $x \in X$ implies strong stability of T and, in the case of a Hilbert space, is equivalent to it.*

Proof. To prove the first part of the theorem we take $x \in X$, $n \in \mathbb{N}$, and $r > 1$. By Lemma 1.11 and the Cauchy-Schwarz inequality we have

$$|\langle T^n x, y \rangle| \leq \frac{r^{n+2}}{2\pi(n + 1)} \int_0^{2\pi} |\langle R^2(re^{i\varphi}, T)x, y \rangle| d\varphi$$

$$\leq \frac{r^{n+2}}{2\pi(n + 1)} \left(\int_0^{2\pi} \|R(re^{i\varphi}, T)x\|^2 d\varphi \right)^{\frac{1}{2}} \left(\int_0^{2\pi} \|R(re^{i\varphi}, T')y\|^2 d\varphi \right)^{\frac{1}{2}}$$

for every $y \in X'$. By (a) and the uniform boundedness principle there exists a constant $M > 0$ such that

$$(r - 1) \int_0^{2\pi} \|R(re^{i\varphi}, T')y\|^2 d\varphi \leq M^2 \|y\|^2 \quad \text{for every } y \in X' \text{ and } r > 1.$$

(The uniform boundedness principle should be applied to the family of operators $S_r : X' \to L^2([0, 2\pi], X')$, $r > 1$, given by $(S_r y)(\varphi) := \sqrt{r - 1} R(re^{i\varphi}, T')y$.) Therefore, we obtain

$$\|T^n x\| \leq \frac{M r^{n+2}}{2\pi(n + 1)(r - 1)} \left((r - 1) \int_0^{2\pi} \|R(re^{i\varphi}, T)x\|^2 d\varphi \right)^{\frac{1}{2}}. \tag{II.15}$$

For $r := 1 + \frac{1}{n+1}$, we obtain $\frac{r^{n+2}}{(n+1)(r-1)} = \left(1 + \frac{1}{n+1}\right)^{n+2} \to e$ as $n \to \infty$, hence $\lim_{n\to\infty} \|T^n x\| = 0$ by (II.15).

Assume now that X is a Hilbert space and T is strongly stable. Then by Lemma 1.11 and Parseval's equality,

$$(r-1) \int_0^{2\pi} \|R(re^{i\varphi}, T)x\|^2 d\varphi = (r-1) \sum_{n=0}^{\infty} \frac{\|T^n x\|^2}{r^{2(n+1)}} = \frac{1}{r+1}(1-s) \sum_{n=0}^{\infty} s^n \|T^n x\|^2$$

for $s := \frac{1}{r^2} < 1$. The right-hand side is, up to the factor $1/(r+1)$, the Abel mean of $\{\|T^n x\|^2\}$. Therefore it converges to zero as $s \to 1$ by the strong stability of T, proving the first part of (a). The second part of (a) follows from Theorem 1.12. $\square$

We now show that the converse implication in Theorem 2.23 does not hold on Banach spaces. We use ideas as in Example 1.14.

Example 2.24. Consider the space l^∞ and the multiplication operator T given by

$$T(x_1, x_2, x_3, \ldots) := (a_1 x_1, a_2 x_2, a_3 x_3, \ldots)$$

for a sequence $\{a_n\}_{n=1}^\infty \subset \{z : |z| < 1\}$ satisfying $\Gamma \subset \overline{\{a_n : n \in \mathbb{N}\}}$. For $|\lambda| > 1$ and $x \in U_{\frac{1}{2}}(1)$ we obtain

$$\|R(\lambda, T)x\| = \sup_{n\in\mathbb{N}} \frac{|x_n|}{|\lambda - a_n|} \geq \frac{1}{2\mathrm{dist}(\lambda, \Gamma)} = \frac{1}{2(|\lambda| - 1)}$$

and therefore

$$(r-1) \int_0^{2\pi} \|R(re^{i\varphi}, T)x\|^2 d\varphi \geq \frac{\pi(r-1)}{2(r-1)^2} = \frac{\pi}{2(r-1)} \to \infty \quad \text{as } r \to 1+.$$

Consider now $X = l^1$ and the multiplication operator S defined by the sequence $\{a_n\}_{n=1}^\infty$. Then S is strongly stable, see Example 2.2 (b). On the other hand, since $S' = T$ and by the considerations above, the second part of (a) in Theorem 1.12 is not fulfilled for an open set in X'.

By Theorem 1.12 we immediately obtain the following characterisation of strongly stable operators on Hilbert spaces.

Corollary 2.25. *Let T be a power bounded operator on a Hilbert space H and $x \in X$. Then $\|T^n x\| \to 0$ if and only if*

$$\lim_{r\to 1+} (r-1) \int_0^{2\pi} \|R(re^{i\varphi}, T)x\|^2 d\varphi = 0. \tag{II.16}$$

In particular, T is strongly stable if and only if (II.16) holds for every $x \in H$ (or only for every x in a dense subset of H).

It remains an open question whether the assertion of Corollary 2.25 holds on arbitrary Banach spaces. More generally, it is not clear what kind of resolvent conditions characterise strong stability of operators on Banach spaces.

3   Weak stability

We now consider stability of operators with respect to the weak operator topology. Surprisingly, this is much more difficult to characterise than the previous concepts.

3.1   Preliminaries

We begin with the definition and some immediate properties of weakly stable operators.

Definition 3.1. Let X be a Banach space. An operator $T \in \mathcal{L}(X)$ is called *weakly stable* if $\lim_{n \to \infty} \langle T^n x, y \rangle = 0$ for every $x \in X$ and $y \in X'$.

Note that by the uniform boundedness principle every weakly stable operator T on a Banach space is power bounded and hence $\sigma(T) \subset \{z : |z| \leq 1\}$. Moreover, the spectral conditions $P_\sigma(T) \cap \Gamma = \emptyset$ and $R_\sigma(T) \cap \Gamma = P_\sigma(T') \cap \Gamma = \emptyset$ are necessary for weak stability.

Example 3.2.   (a) The left and right shifts are weakly stable on the spaces $c_0(\mathbb{Z}, X)$ and $l^p(\mathbb{Z}, X)$ for any Banach space X and $1 < p < \infty$. Note that these operators are isometries and therefore not strongly stable.

(b) Consider $H := l^2$ and the multiplication operator T given by

$$T(x_n)_{n=1}^\infty := (a_n x_n)_{n=1}^\infty$$

for some bounded sequence $(a_n)_{n=1}^\infty$. Then T is weakly stable if and only if $|a_n| < 1$ for every $n \in \mathbb{N}$. However, in this case T is automatically strongly stable.

(c) The situation is different for a multiplication operator on $L^2(\mathbb{R})$ with respect to the Lebesgue measure μ. Let T be defined as $(Tf)(s) := a(s)f(s)$ for some bounded measurable function $a : \mathbb{R} \to \mathbb{R}$. Then T is strongly stable if and only if $|a(s)| < 1$ for almost all s. However, the operator T is weakly stable if and only if $|a(s)| \leq 1$ for almost all s and $\int_c^d a^n(s)ds \to 0$ as $n \to \infty$ for every interval $[c, d] \subset \mathbb{R}$. (Check weak convergence on characteristic functions and use the standard density argument.) This is the case for, e.g., $a(s) = e^{i\alpha s^\gamma}$, $\alpha, \gamma \in \mathbb{R} \setminus \{0\}$.
More examples will be given in Section 5.

We now present a simple condition implying weak stability and begin with the following definition, see, e.g., Furstenberg [92, p. 28].

Definition 3.3. A subsequence $\{n_j\}_{j=1}^\infty$ of $\mathbb{N}$ is called *relatively dense* or *syndetic* if there exists a number $\ell > 0$ such that for every $n \in \mathbb{N}$ the set $\{n, n+1, \ldots, n+\ell\}$ intersects $\{n_j\}_{j=1}^\infty$.

For example, $k\mathbb{N} + m$ for arbitrary natural numbers k and m is a relatively dense subsequence of $\mathbb{N}$.

Theorem 3.4. *Let X be a Banach space and $T \in \mathcal{L}(X)$. Suppose that $\lim_{j\to\infty} T^{n_j} = 0$ weakly for some relatively dense subsequence $\{n_j\}_{j=1}^{\infty}$. Then T is weakly stable.*

Proof. Define $\ell := \sup_{j\in\mathbb{N}}(n_{j+1} - n_j)$ which is finite by assumption and fix $x \in X$ and $y \in X'$. For $n \in \{n_j, \ldots, n_j + \ell\}$ we have

$$\langle T^n x, y \rangle = \langle T^{n-n_j} x, T'^{n_j} y \rangle. \tag{II.17}$$

Note that $T^{n-n_j} x$ belongs to the finite set $\{x, Tx, \ldots, T^{\ell} x\}$. By assumption we have $\lim_{j\to\infty} \langle z, T'^{n_j} y \rangle = 0$ for every $z \in X$, and (II.17) implies $\lim_{n\to\infty} \langle T^n x, y \rangle = 0$. $\qquad\square$

Remark 3.5. We will see later that one cannot replace relative density by density 1, see Section 4.

We now state the following characterisation of weak convergence in terms of (strong) convergence of the Cesàro means of subsequences due to Lin [167], Jones, Kuftinec [136] and Akcoglu, Sucheston [2].

Theorem 3.6. *Let X be a Banach space and $T \in \mathcal{L}(X)$. Consider the following assertions.*

(i) *$T^n x$ converges weakly as $n \to \infty$ for every $x \in X$;*

(ii) *For every $x \in X$ and every increasing sequence $\{n_j\}_{j=1}^{\infty} \subset \mathbb{N}$ with positive lower density, the limit*

$$\lim_{N\to\infty} \frac{1}{N} \sum_{j=1}^{N} T^{n_j} x \text{ exists in norm for every } x \in X; \tag{II.18}$$

(iii) *Property (II.18) holds for every $x \in X$ and every increasing sequence $\{n_j\}_{j=1}^{\infty} \subset \mathbb{N}$.*

Then (i)$\Leftrightarrow$(ii)$\Leftarrow$(iii), and they all are equivalent provided X is a Hilbert space and T is a contraction.

Here, the *lower density* of a sequence $\{n_j\}_{j=1}^{\infty} \subset \mathbb{N}$ is defined by

$$\underline{d} := \liminf_{n\to\infty} \frac{\#\{j : n_j < n\}}{n} \in [0, 1].$$

If the limit on the right-hand side exists, it is called *density* and denoted by d.

Remark 3.7. Assertion (iii) in Theorem 3.6 is called the *Blum–Hanson property* and appears frequently in ergodic theory. For a survey on this property and some recent results and applications to some classical problems in operator theory (e.g., to the quasi-similarity problem) see Müller, Tomilov [192]. In particular, they showed that for power bounded operators on Hilbert spaces the Blum–Hanson property does not follow from weak convergence.

We also add a result of Müller [189] on possible decay of weak orbits.

Theorem 3.8 (Müller). *Let T be an operator on a Banach space X with $r(T) \geq 1$ and $\{a_n\}_{n=1}^{\infty}$ be a positive sequence satisfying $a_n \to 0$. Then there exist $x \in X$, $y \in X'$ and an increasing sequence $\{n_j\}_{j=1}^{\infty} \subset \mathbb{N}$ such that*

$$\mathrm{Re}\,\langle T^{n_j} x, y \rangle \geq a_j \qquad \forall j \in \mathbb{N}. \tag{II.19}$$

Surprisingly, inequality (II.19) does not hold for all $n \in \mathbb{N}$ in general, i.e., the weak version of Theorem 2.8 is not true. For an example and many other phenomena see Müller [189] and [191, Section V.39]. However, for weakly stable operators one can choose $n_j := j$, see Badea, Müller [12].

3.2 Contractions on Hilbert spaces

In this subsection we present classical decomposition theorems for contractions on Hilbert spaces having direct connection to weak stability.

The first theorem is due to Sz.-Nagy, Foiaş [237], see also [238].

Theorem 3.9 (Sz.-Nagy, Foiaş, 1960). *Let T be a contraction on a Hilbert space H. Then H is the orthogonal sum of two T- and T^*-invariant subspaces H_1 and H_2 such that*

(a) *H_1 is the maximal subspace on which the restriction of T is unitary;*

(b) *the restrictions of T and T^* to H_2 are weakly stable.*

We follow the proof given by Foguel [86].

Proof. Define

$$H_1 := \{x \in H : \|T^n x\| = \|T^{*n} x\| = \|x\| \quad \text{for all } n \in \mathbb{N}\}.$$

We first prove that for every $x \in H_1$ and $n \in \mathbb{N}$ one has $T^{*n} T^n x = T^n T^{*n} x = x$. If $x \in H_1$, then $\|x\|^2 = \langle T^n x, T^n x \rangle = \langle T^{*n} T^n x, x \rangle \leq \|T^{*n} T^n x\| \|x\| \leq \|x\|^2$ holds. Therefore, by the equality in the Cauchy-Schwarz inequality and positivity of $\|x\|^2$ we have $T^{*n} T^n x = x$. Analogously one shows $T^n T^{*n} x = x$. On the other hand, every x with these two properties belongs to H_1. So we proved that

$$H_1 = \{x \in H : T^{*n} T^n x = T^n T^{*n} x = x \quad \text{for all } n \in \mathbb{N}\} \tag{II.20}$$

is the maximal (closed) subspace on which T is unitary. The T- and T^*-invariance of H_1 follows from the definition of H_1 and the fact that $T^* T = T T^*$ on H_1.

To prove (b) take $x \in H_2 := H_1^{\perp}$. Note that H_2 is T- and T^*-invariant since H_1 is. Suppose that $T^n x$ does not converge weakly to zero as $n \to \infty$. This means that there exists $\varepsilon > 0$, $y \in H_2$ and a subsequence $\{n_j\}_{j=1}^{\infty}$ such that $|\langle T^{n_j} x, y \rangle| \geq \varepsilon$ for every $j \in \mathbb{N}$. On the other hand, $\{T^{n_j} x\}_{j=1}^{\infty}$ contains a weakly converging subsequence which we again denote by $\{T^{n_j} x\}_{j=1}^{\infty}$, and its limit by x_0.

Since H_2 is T-invariant and closed, x_0 belongs to H_2. To achieve a contradiction we show below that actually $x_0 = 0$.

For a fixed $k \in \mathbb{N}$ we have

$$\|T^{*k}T^kT^nx - T^nx\|^2 = \|T^{*k}T^{k+n}x\|^2 - 2\langle T^{*k}T^{k+n}x, T^nx\rangle + \|T^nx\|^2$$
$$\leq \|T^{k+n}x\|^2 - 2\|T^{k+n}x\|^2 + \|T^nx\|^2 = \|T^nx\|^2 - \|T^{k+n}x\|^2.$$

The right-hand side converges to zero as $n \to \infty$ since the sequence $\{\|T^nx\|\}_{n=1}^{\infty}$ is monotone decreasing and therefore convergent. So $\|T^{*k}T^kT^nx - T^nx\| \to 0$ as $n \to \infty$.

We now return to the above subsequence $\{T^{n_j}x\}_{j=1}^{\infty}$ converging weakly to x_0. Then $T^{*k}T^kT^{n_j}x \to T^{*k}T^kx_0$ weakly. On the other hand, by the considerations above, we have $T^{*k}T^kT^{n_j}x \to x_0$ weakly and therefore $T^{*k}T^kx_0 = x_0$. One shows analogously that $T^kT^{*k}x_0 = x_0$ and hence $x_0 \in H_1$. Since $H_1 \cap H_2 = \{0\}$, we obtain $x_0 = 0$, the desired contradiction.

Analogously, the powers of the restriction of T^* to H_2 converge weakly to zero. $\square$

Remark 3.10. The restriction of T to the subspace H_2 in Theorem 3.9 is *completely non-unitary* (c.n.u. for short), i.e., there is no non-trivial subspace of H_2 on which the restriction of T becomes unitary. In other words, Theorem 3.9 states that every Hilbert space contraction can be decomposed into a unitary and a c.n.u. part, and the c.n.u. part is weakly stable.

For a systematic study of completely non-unitary operators as well as an alternative proof of Theorem 3.9 using unitary dilation theory see the monograph of Sz.-Nagy and Foiaş [238].

On the other hand, we have the following decomposition into the weakly stable and the weakly unstable part due to Foguel [86]. We give a simplified proof.

Theorem 3.11 (Foguel, 1963). *Let T be a contraction on a Hilbert space H. Then*

$$W := \{x \in H : \lim_{n \to \infty} \langle T^nx, x\rangle = 0\}$$

coincides with

$$\{x \in H : \lim_{n \to \infty} T^nx = 0 \text{ weakly }\} = \{x \in H : \lim_{n \to \infty} T^{*n}x = 0 \text{ weakly }\}$$

and is a closed T- and T^*-invariant subspace of H, and the restriction of T to $W^{\perp}$ is unitary.

Proof. We first take $x \in W$ and show that $T^nx \to 0$ weakly. By Theorem 3.9 we may assume that $x \in H_1$. Take $S := \overline{\lim}\{T^nx : n = 0, 1, 2 \ldots\}$. Since for all $y \in S^{\perp}$ we have $\langle T^nx, y\rangle = 0$ for all n, it is enough to show that $\lim_{n \to \infty}\langle T^nx, y\rangle = 0$ for all $y \in S$. For $y := T^kx$ we obtain

$$\langle T^nx, y\rangle = \langle T^{*k}T^nx, x\rangle = \langle T^{n-k}x, x\rangle \to 0 \quad \text{for } k \leq n \to \infty,$$

where we used that the restriction of T to H_1 is unitary. From the density of $\operatorname{lin}\{T^n x : n = 0, 1, 2, \ldots\}$ in S, it follows that $\lim_{n\to\infty}\langle T^n x, y\rangle = 0$ for every $y \in S$ and therefore $\lim_{n\to\infty} T^n x = 0$ weakly. Analogously, one shows that $\lim_{n\to\infty} T^{*n} x = 0$ weakly. The converse implication, the closedness and the invariance of W are clear.

The last assertion of the theorem follows directly from Theorem 3.9. $\square$

Combining Theorem 3.9 and Theorem 3.11 we obtain the following decomposition.

Theorem 3.12. *Let T be a contraction on a Hilbert space H. Then H is the orthogonal sum of three closed T- and T^*-invariant subspaces H_1, H_2 and H_3 such that the corresponding restrictions T_1, T_2 and T_3 satisfy*

1. *T_1 is unitary and has no non-zero weakly stable orbit;*

2. *T_2 is unitary and weakly stable;*

3. *T_3 is completely non-unitary and weakly stable.*

The above theorem shows that a characterisation of weak stability for unitary operators is of special importance. For more aspects of this problem see Section IV.1.

3.3   Characterisation via resolvent

To close this section we give a resolvent characterisation of weak stability being a discrete analogue of a result for C_0-semigroups due to Chill, Tomilov [49], see Subsection III.4.3.

Theorem 3.13. *Let T be an operator on a Banach space X with $r(T) \le 1$ and take $x \in X$ and $y \in X'$. Consider the following assertions.*

(a)
$$\int_1^2 \int_0^{2\pi} |\langle R^2(re^{i\varphi}, T)x, y\rangle|\, d\varphi\, dr < \infty.$$

(b)
$$\lim_{r\to 1+} (r-1) \int_0^{2\pi} |\langle R^2(re^{i\varphi}, T)x, y\rangle|\, d\varphi = 0.$$

(c)
$$\lim_{n\to\infty} \langle T^n x, y\rangle = 0.$$

Then (a)$\Rightarrow$(b)$\Rightarrow$(c). In particular, if (a) or (b) hold for all $x \in X$ and $y \in X'$, then T is weakly stable.

Note that (a) means that the function $z \mapsto \langle R^2(z, T)x, y\rangle$ belongs to the Bergman space $A^1(\{z \in \mathbb{C} : 1 < |z| < 2\})$. (See Hedenmalm, Korenblum, Zhu [124] or Bergman [32] for basic information on Bergman spaces.)

Proof. We first prove (a)⇒(b). The function $g : \mathbb{D} = \{z \in \mathbb{C} : |z| < 1\} \to \mathbb{C}$ defined by

$$g(z) := \begin{cases} \langle R^2(z^{-1}, T)x, y \rangle & \text{if } 0 < |z| < 1, \\ 0 & \text{if } z = 0 \end{cases}$$

is holomorphic in $\mathbb{D}$. (Holomorphy of g in $z = 0$ is a consequence of the Neumann representation of the resolvent.) It follows from the classical theory of Hardy spaces on the unit disc, see e.g. Rosenblum, Rovnyak [223, Theorem 2.6], that the function

$$r \mapsto \int_0^{2\pi} |g(re^{i\varphi})|\, d\varphi$$

is monotone increasing on $(0, 1)$ which implies that

$$f(r) := \int_0^{2\pi} |\langle R^2(re^{i\varphi}, T)x, y \rangle|\, d\varphi$$

increases as $r \to 1+$. Assume that (b) fails. Then there exists a decreasing sequence $\{r_n\}_{n=1}^\infty$ converging to 1 and a constant $c > 0$ such that

$$(r_n - 1)f(r_n) \geq c \quad \text{for every } n \in \mathbb{N}. \tag{II.21}$$

Take now n and m with $r_n - 1 \leq (r_m - 1)/2$. The monotonicity of f and (II.21) imply

$$\int_{r_n}^{r_m} f(r)\, dr \geq f(r_m)(r_m - r_n) \geq \frac{c}{r_m - 1}(r_m - r_n) = c\left(1 - \frac{r_n - 1}{r_m - 1}\right) \geq \frac{c}{2}$$

contradicting (a).

We now assume (b). By formula (II.2) we have

$$\langle T^n x, y \rangle \leq \frac{r^{n+2}}{2\pi(r - 1)(n + 1)}(r - 1)\int_0^{2\pi} |\langle R^2(re^{i\varphi}, T)x, y \rangle|\, d\varphi$$

for all $n \in \mathbb{N}$ and $r > 1$. Taking $r := 1 + \frac{1}{n+1}$ we obtain by (b) that $\langle T^n x, y \rangle \to 0$ as $n \to \infty$. $\qquad\square$

Remark 3.14. As mentioned above, a (simple) necessary and sufficient resolvent condition for weak stability is still unknown. In particular, it is not clear whether condition (b) in the above theorem is necessary.

4   Almost weak stability

In this section we consider a stability concept which is analogous to weak mixing in ergodic theory (see Halmos [118]). This notion is weaker and much easier to investigate than weak stability. In fact, a complete characterisation through spectral or resolvent conditions is available as we show below. In large parts we modify and extend the treatment in Eisner, Farkas, Nagel, Sereny [67] in the continuous case.

4.1 Characterisation

The main result of this section is the following list of equivalent properties.

Theorem 4.1. *Let T be an operator on a Banach space X having relatively weakly compact orbits. Then the following assertions are equivalent.*

(i) $0 \in \overline{\{T^n x : n \in \mathbb{N}\}}^\sigma$ *for every* $x \in X$.

(i') $0 \in \overline{\{T^n : n \in \mathbb{N}\}}^{\mathcal{L}_\sigma}$.

(ii) *For every $x \in X$ there exists a subsequence $\{n_j\}_{j=1}^\infty \subset \mathbb{N}$ such that*
$$\lim_{j \to \infty} T^{n_j} x = 0 \text{ weakly.}$$

(iii) *For every $x \in X$ there exists a subsequence $\{n_j\}_{j=1}^\infty \subset \mathbb{N}$ with density 1 such that* $\lim_{j \to \infty} T^{n_j} x = 0$ *weakly.*

(iv) $\displaystyle \lim_{n \to \infty} \frac{1}{n+1} \sum_{k=0}^{n} |\langle T^k x, y \rangle| = 0$ *for all* $x \in X$ *and* $y \in X'$.

(v) $\displaystyle \lim_{r \to 1+} (r - 1) \int_0^{2\pi} |\langle R(re^{i\varphi}, T)x, y \rangle|^2 \, d\varphi = 0$ *for all* $x \in X$ *and* $y \in X'$.

(vi) $\displaystyle \lim_{r \to 1+} (r - 1) R(re^{i\varphi}, T)x = 0$ *for all* $x \in X$ *and* $0 \le \varphi < 2\pi$.

(vii) $P_\sigma(T) \cap \Gamma = \emptyset$, *i.e., T has no eigenvalues on the unit circle.*

In addition, if X' is separable, then the following conditions are equivalent to the conditions above.

(ii*) *There exists a subsequence $\{n_j\}_{j=1}^\infty \subset \mathbb{N}$ such that* $\displaystyle \lim_{j \to \infty} T^{n_j} = 0$ *in the weak operator topology.*

(iii*) *There exists a subsequence $\{n_j\}_{j=1}^\infty \subset \mathbb{N}$ with density 1 such that* $\displaystyle \lim_{j \to \infty} T^{n_j} = 0$ *in the weak operator topology.*

Recall that the density of a sequence $\{n_j\}_{j=1}^\infty \subset \mathbb{N}$ was defined after Theorem 3.6.

The following elementary lemma (see, e.g., Petersen [212, p. 65] or Lemma III.5.2 below in the continuous case) will be needed in the proof of Theorem 4.1.

Lemma 4.2 (Koopman–von Neumann, 1932). *For a bounded sequence $\{a_n\}_{n=1}^\infty \subset [0, \infty)$ the following assertions are equivalent.*

(a) $\displaystyle \lim_{n \to \infty} \frac{1}{n} \sum_{k=1}^{n} a_k = 0$.

(b) *There exists a subsequence $\{n_j\}_{j=1}^\infty$ of $\mathbb{N}$ with density 1 such that* $\lim_{j \to \infty} a_{n_j} = 0$.

Proof of Theorem 4.1. The implications (i')⇒(i) and (ii)⇒(vii) are trivial.

(i)⇒(ii) follows from the equivalence of weak compactness and weak sequential compactness in Banach spaces (see Theorem I.1.1).

The implication (vii)⇒(i') is a consequence of Theorem I.1.15 and the construction in its proof. Therefore, we already proved the equivalences (i)⇔(i')⇔(ii) ⇔(vii).

(vi)⇒(vii): Assume that there exists $0 \neq x \in X$ and $\varphi \in [0, 2\pi)$ such that $Tx = e^{i\varphi}x$. Then we have $\|R(re^{i\varphi}, T)x\| = (r-1)^{-1}\|x\|$ and (vi) does not hold.

The converse implication follows from Theorem 2.8 and the fact that for every $0 \leq \varphi < 2\pi$ the operator $e^{i\varphi}T$ has relatively weakly compact orbits as well. So (vi)⇔(vii).

(i')⇒(iii): Take $S := \overline{\{T^n x : n \geq 0\}}^{\mathcal{L}_\sigma(X)} \subset \mathcal{L}(X)$ with the usual multiplication and the weak operator topology. It becomes a compact semitopological semigroup (for the definition and basic properties of compact semitopological semigroups see Subsection I.1.4). By (i') we have $0 \in S$. Define the operator $\tilde{T} : C(S) \to C(S)$ by

$$(\tilde{T}f)(R) := f(TR), \quad f \in C(S), \ R \in S.$$

Note that $\tilde{T}$ is a contraction on $C(S)$.

By Example I.1.7 (c) the set $\{f(T^n \cdot) : n \geq 0\}$ is relatively weakly compact in $C(S)$ for every $f \in C(S)$. It means that every set $\{\tilde{T}^n f : n \geq 0\}$ is relatively weakly compact, i.e., $\tilde{T}$ has relatively weakly compact orbits.

Denote by $\tilde{P}$ the mean ergodic projection of $\tilde{T}$. We have $\text{Fix}(\tilde{T}) = \langle \mathbf{1} \rangle$. Indeed, for $f \in \text{Fix}(\tilde{T})$ one has $f(T^n I) = f(I)$ for all $n \geq 0$ and therefore f must be constant. Hence $\tilde{P}f$ is constant for every $f \in C(S)$. By definition of the ergodic projection

$$(\tilde{P}f)(0) = \lim_{n\to\infty} \frac{1}{n+1}\sum_{k=0}^{n}(\tilde{T}^k f)(0) = f(0). \tag{II.22}$$

Thus we have

$$(\tilde{P}f)(R) = f(0) \cdot \mathbf{1}, \quad f \in C(S), \ R \in S. \tag{II.23}$$

Take now $x \in X$. By Theorem I.1.5 and its proof (see Dunford, Schwartz [63, p. 434]), the weak topology on the orbit $\{T^n x : n \geq 0\}$ is metrisable and coincides with the topology induced by some sequence $\{y_n\}_{n=1}^{\infty} \subset X' \setminus \{0\}$. Consider $f_x \in C(S)$ defined by

$$f_x(R) := \sum_{n\in\mathbb{N}} \frac{1}{2^n}\left|\left\langle Rx, \frac{y_n}{\|y_n\|}\right\rangle\right|, \quad R \in S.$$

By (II.23) we obtain

$$0 = \lim_{n\to\infty} \frac{1}{n+1}\sum_{k=0}^{n}(\tilde{T}^k f_x)(I) = \lim_{n\to\infty} \frac{1}{n+1}\sum_{k=0}^{n} f_x(T^k).$$

Lemma 4.2 applied to the sequence $\{f_x(T^n I)\}_{n=0}^\infty \subset \mathbb{R}_+$ yields a subsequence $(n_j)_{j=1}^\infty$ of $\mathbb{N}$ with density 1 such that

$$\lim_{j\to\infty} f_x(T^{n_j}) = 0.$$

By definition of f_x and by the fact that the weak topology on the orbit is induced by $\{y_n\}_{n=1}^\infty$ we have that

$$\lim_{j\to\infty} T^{n_j} x = 0 \quad \text{weakly},$$

and (iii) is proved.

(iii)$\Rightarrow$(iv) follows directly from Lemma 4.2.

(iv)$\Rightarrow$(vii) is clear.

(iv)$\Leftrightarrow$(v): We note first that the set $\{T^n : n \geq 0\}$ is bounded in $\mathcal{L}(X)$. Take $x \in X$, $y \in X'$ and let $r > 1$. By (II.2) and the Parseval's equality we have

$$\int_0^{2\pi} |\langle R(re^{i\varphi}, T)x, y\rangle|^2 \, d\varphi = 2\pi \sum_{n=0}^\infty \frac{|\langle T^n x, y\rangle|^2}{r^{2n+2}}.$$

We obtain by the equivalence of Abel and Cesàro limits (see Lemma I.2.6) that

$$\lim_{r\to 1+} (r-1) \int_0^{2\pi} |\langle R(re^{i\varphi}, T)x, y\rangle|^2 \, d\varphi = \pi \lim_{n\to\infty} \frac{1}{n+1} \sum_{k=0}^n |\langle T^k x, y\rangle|^2. \quad \text{(II.24)}$$

Note that for a bounded sequence $\{a_n\}_{n=1}^\infty \subset \mathbb{R}_+$ with $C := \sup_{n\in\mathbb{N}} a_n$ we have

$$\left(\frac{1}{C(n+1)} \sum_{k=0}^n a_k^2\right)^2 \leq \left(\frac{1}{n+1} \sum_{k=0}^n a_k\right)^2 \leq \frac{1}{n+1} \sum_{k=0}^n a_k^2$$

for every $n \in \mathbb{N}$, where for the second part we used the Cauchy-Schwarz inequality. This together with (II.24) gives the equivalence of (iv) and (v).

For the additional part of the assertion suppose X' to be separable. Then so is X, and we can take dense subsets $\{x_n \neq 0 : n \in \mathbb{N}\} \subseteq X$ and $\{y_m \neq 0 : m \in \mathbb{N}\} \subseteq X'$. Consider the functions

$$f_{n,m} : S \to \mathbb{R}, \qquad f_{n,m}(R) := \left|\left\langle R\frac{x_n}{\|x_n\|}, \frac{y_m}{\|y_m\|}\right\rangle\right|, \quad n, m \in \mathbb{N},$$

which are continuous and uniformly bounded in $n, m \in \mathbb{N}$. Define the function

$$f : S \to \mathbb{R}, \qquad f(R) := \sum_{n,m\in\mathbb{N}} \frac{1}{2^{n+m}} f_{n,m}(R).$$

Then clearly $f \in C(S)$. Thus, as in the proof of the implication (i')⇒(iii), i.e., using (II.22) we obtain

$$\lim_{n\to\infty} \frac{1}{n+1} \sum_{k=0}^{n} f(T^n I) = 0.$$

Hence, Lemma 4.2 applied to the bounded sequence $\{f(T^n)\}_{n=0}^{\infty} \subset \mathbb{R}_+$ yields the existence of a subsequence $\{n_j\}_{j=1}^{\infty}$ of $\mathbb{N}$ with density 1 such that $\lim_{j\to\infty} f(T^{n_j}) = 0$. In particular, $\lim_{j\to\infty} |\langle T^{n_j} x_n, y_m\rangle| = 0$ for all $n, m \in \mathbb{N}$, which, together with the boundedness of $\{T^n\}_{n=1}^{\infty}$, proves the implication (i')⇒(iii*). The implications (iii*)⇒(ii*)⇒(ii') are straightforward, hence the proof is complete.   □

The above theorem shows that the property

"no eigenvalues of T on the unit circle"

implies properties like (iii) on the asymptotic behaviour of the orbits of T. Motivated by this we introduce the following terminology.

Definition 4.3. We call an operator on a Banach space with relatively weakly compact orbits *almost weakly stable* if it satisfies condition (iii) in Theorem 4.1.

Historical remark 4.4. Theorem 4.1 and especially the implication (vii)⇒(iii) has a long history. It goes back to the origin of ergodic theory and von Neumann's spectral mixing theorem for flows, see Halmos [118], Mixing Theorem, p. 39. This has been generalised to operators on Banach spaces by many authors, see, e.g., Nagel [195], Jones, Lin [137, 138] and Krengel [154], pp. 108–110. Note that the equivalence (vii)⇔(iv) for contractions on Hilbert spaces also follows from the so-called generalised Wiener theorem, see Goldstein [102].

For a continuous analogue of the above characterisation see Theorem III.5.1.

Remark 4.5. We emphasise that the conditions appearing in Theorem 4.1 are of quite different nature. Conditions (i)–(iv), (ii*) and (iii*) describe the behaviour of the powers of T, while conditions (v)–(vii) consider the spectrum and the resolvent of T in a neighbourhood of the unit circle. Among them condition (vii) apparently is the simplest to verify. Note that one can also add the equivalent condition

(iv') $\displaystyle\lim_{n\to\infty} \sup_{y\in X', \|y\|\leq 1} \frac{1}{n+1} \sum_{k=0}^{n} |\langle T^k x, y\rangle| = 0$ for every $x \in X$,

see Jones, Lin [137, 138].

Remark 4.6. The equivalence (i')⇔(v) is a weak analogue of the characterisation of strong stability given in Corollary 2.25.

Remark 4.7. Theorem 3.4 leads to an interesting algebraic property of the sequence $\{n_j\}_{j=1}^{\infty}$ appearing in condition (iii). Indeed, if $\{n_j\}_{j=1}^{\infty}$ contains a relatively dense subsequence, then the operator T is weakly stable. Reformulating this, we obtain the following: Let T be almost weakly but not weakly stable. Then the sequence $\{n_j\}_{j=1}^\infty$ in (iii) does not contain any relatively dense set (for example, it does not contain any equidistant sequence).

Using Theorem 4.1 one can reformulate the Jacobs–Glicksberg–de Leeuw decomposition (see Theorem I.1.15) as follows.

Theorem 4.8 (Jacobs–Glicksberg–de Leeuw decomposition, extended version). *Let X be a Banach space and let $T \in \mathcal{L}(X)$ have relatively weakly compact orbits. Then $X = X_r \oplus X_s$, where*

$$X_r := \overline{\lin}\{x \in X : Tx = \gamma x \text{ for some } \gamma \in \Gamma\},$$
$$X_s := \{x \in X : \lim_{j \to \infty} T^{n_j} x = 0 \text{ weakly for some subsequence } \{n_j\}_{j=1}^\infty$$
$$\text{with density } 1\}.$$

One also can formulate Theorem 4.1 for single orbits. This is the following result partially due to Jan van Neerven (oral communication) in the continuous case.

Corollary 4.9. *Let T be an operator on a Banach space X and $x \in X$. Assume that the orbit $\{T^n x : n = 0, 1, 2, \ldots\}$ is relatively weakly compact in X and the restriction of T to $\overline{\lin}\{T^n x : n = 0, 1, 2, \ldots\}$ is power bounded. Then there is a holomorphic continuation of the function $R(\cdot, T)x$ to $\{\lambda : |\lambda| > 1\}$ denoted by $R_x(\cdot)$ and the following assertions are equivalent.*

(i) $0 \in \overline{\{T^n x : n \in \mathbb{N}\}}^\sigma$.

(ii) *There exists a subsequence $\{n_j\}_{j=1}^\infty \subset \mathbb{N}$ such that $\lim_{j \to \infty} T^{n_j} x = 0$ weakly.*

(iii) *There exists a subsequence $\{n_j\}_{j=1}^\infty \subset \mathbb{N}$ with density 1 such that $\lim_{j \to \infty} T^{n_j} x = 0$ weakly.*

(iv) $\displaystyle \lim_{n \to \infty} \frac{1}{n+1} \sum_{k=0}^n |\langle T^k x, y \rangle| = 0$ *for all $y \in X'$.*

(v) $\displaystyle \lim_{r \to 1+} (r-1) \int_0^{2\pi} |\langle R_x(re^{i\varphi}), y \rangle|^2 \, d\varphi = 0$ *for all $y \in X'$.*

(vi) $\displaystyle \lim_{r \to 1+} (r-1) R_x(re^{i\varphi}) = 0$ *for all $0 \le \varphi < 2\pi$.*

(vii) *The restriction of T on $\overline{\lin}\{T^n x : n \in \mathbb{N} \cup \{0\}\}$ has no unimodular eigenvalue.*

Proof. For the first part of the theorem we just define

$$R_x(\lambda) := \sum_{n=0}^\infty \frac{T^n x}{\lambda^{n+1}} \quad \text{whenever } |\lambda| > 1.$$

This implies the representation

$$T^n x = \frac{r^{n+1}}{2\pi} \int_0^{2\pi} e^{(n+1)i\varphi} R_x(re^{i\varphi}) \, d\varphi \quad \text{for all } n \in \mathbb{N}.$$

Denote now by Z the closed linear span of the orbit $\{T^n x : n = 0, 1, 2, \ldots\}$. Then Z is a T-invariant closed subspace of X and we can restrict T to it. The restriction, which we denote by T_Z, has relatively weakly compact orbits by Lemma I.1.6. The equivalence of the assertions follows from the canonical decomposition $X' = Z' \oplus Z^0$ with $Z^0 := \{y \in X' : \langle z, y \rangle = 0 \text{ for all } z \in Z\}$ and Theorem 4.1 applied to the restricted operator. $\qquad\qquad\square$

4.2 Concrete example

As we will see from the abstract examples in the next section, almost weak stability does not imply weak stability. In this subsection we present a concrete example of a (positive) operator being almost weakly but not weakly stable. (For definitions and basic theory of positive operators we refer to Schaefer [227].)

Example 4.10. Consider the operator $T_0(1)$ from Example III.5.9 below being a positive operator on the Banach lattice $C(\Omega)$ for some $\Omega \subset \mathbb{C}$. The relative weak compactness of the orbits follows from the relative weak compactness of the semigroup $T_0(\cdot)$. The almost weak stability of $T_0(1)$ is a consequence of $P_\sigma(T_0(1)) \cap \Gamma = \emptyset$ and Theorem 4.1. Further, since the semigroup $T_0(\cdot)$ is not weakly stable and $\mathbb{N}$ is a relatively dense set in $\mathbb{R}_+$, it follows from Theorem III.4.4 that the operator $T_0(1)$ is not weakly stable.

So we proved the following result.

Theorem 4.11. *There is a locally compact space Ω and a positive contraction T on $C_0(\Omega)$ which is almost weakly but not weakly stable.*

Note that for positive operators on $C(K)$ and L^1-spaces, almost weak stability does imply weak stability and even a stronger stability property, see Chill, Tomilov [50] as well as Groh, Neubrander [112, Theorem. 3.2] for the continuous case.

Theorem 4.12. *For a power bounded, positive, and mean ergodic operator T on a Banach lattice X, the following assertions hold.*

(i) *If $X \cong L^1(\Omega, \mu)$, then $P\sigma(T') \cap \Gamma = \emptyset$ is equivalent to* strong stability *of T.*

(ii) *If $X \cong C(K)$, K compact, then $P\sigma(T') \cap \Gamma = \emptyset$ is equivalent to* uniform exponential stability *of T.*

5 Category theorems

By proving category results, we obtain abstract examples of almost weakly but not weakly stable operators and show that the difference between these two concepts (at least for operators on Hilbert spaces) is dramatic. More precisely, we show that a "typical" (in the sense of Baire) contraction as well as a "typical" isometric or unitary operator on a separable Hilbert space is almost weakly but not weakly stable. This gives an operator–theoretic analogue to the classical theorems of Halmos and Rohlin from ergodic theory stating that a "typical" measure preserving transformation is weakly but not strongly mixing, see Halmos [118, pp. 77–80] or the original papers by Halmos [116] and Rohlin [221], see also Section IV.2. We follow in this section Eisner, Serény [69] and assume the underlying Hilbert space H to be separable and infinite-dimensional.

5.1 Unitary operators

Denote the set of all unitary operators on H by $\mathcal{U}$. The following density result for periodic operators is a first step in our construction.

Proposition 5.1. *For every $n \in \mathbb{N}$ the set of all periodic unitary operators with period greater than n is dense in $\mathcal{U}$ endowed with the operator norm topology.*

Proof. Take $U \in \mathcal{U}$, $N \in \mathbb{N}$ and $\varepsilon > 0$. By the spectral theorem H is isomorphic to $L^2(\Omega, \mu)$ for some finite measure μ on a locally compact space Ω and U is unitarily equivalent to the multiplication operator $\tilde{U}$ with

$$(\tilde{U}f)(\omega) = \varphi(\omega)f(\omega) \quad \text{for almost all } \omega \in \Omega,$$

for some measurable $\varphi : \Omega \to \Gamma$.

We approximate the operator $\tilde{U}$ as follows. Consider the set

$$\Gamma_N := \left\{ e^{2\pi i \cdot \frac{p}{q}} : p, q \in \mathbb{N} \text{ relatively prime}, q > N \right\}$$

which is dense in Γ. Take a finite set $\{\alpha_j\}_{j=1}^n \subset \Gamma_N$ with $\alpha_1 = \alpha_n$ such that $\arg(\alpha_{j-1}) < \arg(\alpha_j)$ and $|\alpha_j - \alpha_{j-1}| < \varepsilon$ hold for all $2 \leq j \leq n$. Define

$$\psi(\omega) := \alpha_{j-1}, \quad \forall \omega \in \varphi^{-1}(\{z \in \Gamma : \arg(\alpha_{j-1}) \leq \arg(z) < \arg(\alpha_j)\}),$$

and denote by $\tilde{P}$ the multiplication operator with ψ. The operator $\tilde{P}$ is periodic with period greater than N and satisfies

$$\|\tilde{U} - \tilde{P}\| = \sup_{\omega \in \Omega} |\varphi(\omega) - \psi(\omega)| \leq \varepsilon,$$

hence the proposition is proved. □

For the second step we need the following lemma.

Lemma 5.2. *Let H be a separable infinite-dimensional Hilbert space. Then there exists a sequence $\{T_n\}_{n=1}^\infty$ of almost weakly stable unitary operators satisfying $\lim_{n\to\infty} \|T_n - I\| = 0$.*

Proof. It suffices to prove the result for $H = L^2(\mathbb{R})$.

Take $n \in \mathbb{N}$ and define $\tilde{T}_n$ on $L^2(\mathbb{R})$ by

$$(\tilde{T}_n f)(s) := e^{\frac{iq(s)}{n}} f(s), \quad s \in \mathbb{R}, \quad f \in L^2(\mathbb{R}),$$

where $q : \mathbb{R} \to [0,1]$ is strictly monotone. Then all $\tilde{T}_n$ are almost weakly stable by Theorem 4.1, and they approximate I since

$$\|\tilde{T}_n - I\| = \sup_{s \in \mathbb{R}} |e^{\frac{iq(s)}{n}} - 1| \leq |e^{\frac{i}{n}} - 1| \to 0, \quad n \to \infty.$$

The lemma is proved. $\square$

We now introduce the *strong* (operator) topology* which is induced by the family of seminorms $p_x(T) := \sqrt{\|Tx\|^2 + \|T^*x\|^2}$, $x \in H$. We note that convergence in this topology corresponds to strong convergence of operators and their adjoints. For properties and further information on this topology we refer to Takesaki [239, p. 68].

In the following we consider the space $\mathcal{U}$ of all unitary operators on H endowed with this strong* operator topology. Note that $\mathcal{U}$ is a complete metric space with respect to the metric given by

$$d(U, V) := \sum_{j=1}^\infty \frac{\|Ux_j - Vx_j\| + \|U^*x_j - V^*x_j\|}{2^j \|x_j\|} \quad \text{for } U, V \in \mathcal{U},$$

and $\{x_j\}_{j=1}^\infty$ some dense subset of $H \setminus \{0\}$. Further we denote by $\mathcal{S}_\mathcal{U}$ the set of all weakly stable unitary operators on H and by $\mathcal{W}_\mathcal{U}$ the set of all almost weakly stable unitary operators on H.

We now show the following density property for $\mathcal{W}_\mathcal{U}$.

Proposition 5.3. *The set $\mathcal{W}_\mathcal{U}$ of all almost weakly stable unitary operators is dense in $\mathcal{U}$.*

Proof. By Proposition 5.1 it is enough to approximate periodic unitary operators by almost weakly stable unitary operators. Let U be a periodic unitary operator and let N be its period. Take $\varepsilon > 0$, $n \in \mathbb{N}$ and $x_1, \ldots, x_n \in H \setminus \{0\}$. We have to find an almost weakly stable unitary operator T with $\|Ux_j - Tx_j\| < \varepsilon$ and $\|U^*x_j - T^*x_j\| < \varepsilon$ for all $j = 1, \ldots, n$.

Since $U^N = I$ we have $\sigma(U) \subset \left\{1, e^{\frac{2\pi i}{N}}, \ldots, e^{\frac{2\pi(N-1)i}{N}}\right\}$, and the orthogonal decomposition

$$H = \ker(I - U) \oplus \ker(e^{\frac{2\pi i}{N}} I - U) \oplus \ldots \oplus \ker(e^{\frac{2\pi(N-1)i}{N}} I - U) \tag{II.25}$$

holds.

Assume first that $x_1, \ldots, x_n$ are orthogonal eigenvectors of U.

In order to use Lemma 5.2 we first construct a periodic unitary operator S satisfying $Ux_j = Sx_j$ for all $j = 1, \ldots, n$ and having infinite-dimensional eigenspaces only. For this purpose define the n-dimensional U- and U^*-invariant subspace $H_0 := \lim\{x_j\}_{j=1}^n$ and the operator S_0 on H_0 as the restriction of U to H_0. Decompose H as an orthogonal sum

$$H = \bigoplus_{k=0}^{\infty} H_k \quad \text{with} \dim H_k = \dim H_0 \ \text{for all} \ k \in \mathbb{N}.$$

Denote by P_k an isomorphism from H_k to H_0 for every k. Define now $S_k := P_k^{-1} U P_k$ on each H_k as a copy of $U|_{H_0}$ and consider $S := \bigoplus_{k=0}^{\infty} S_k$ on H.

The operator S is unitary and periodic with period being a divisor of N. So a decomposition analogous to (II.25) is valid for S. Moreover, $Ux_j = Sx_j$ and $U^*x_j = S^*x_j$ hold for all $j = 1, \ldots, n$ and the eigenspaces of S are infinite dimensional. Denote by F_j the eigenspace of S containing x_j and by λ_j the corresponding eigenvalue. By Lemma 5.2, for every $j = 1, \ldots, n$ there exists an almost weakly stable unitary operator T_j on F_j satisfying $\|T_j - S_{|F_j}\| = \|T_j - \lambda_j I\| < \varepsilon$. Finally, we define the desired operator T as T_j on F_j for every $j = 1 \ldots, n$ and extend it linearly to H.

Let now $x_1, \ldots, x_n \in H$ be arbitrary and take an orthonormal basis of eigenvalues $\{y_k\}_{k=1}^{\infty}$. Then there exists $K \in \mathbb{N}$ such that $x_j = \sum_{k=1}^K a_{jk} y_k + o_j$ with $\|o_j\| < \frac{\varepsilon}{4}$ for every $j = 1, \ldots, n$. By the arguments above applied to $y_1, \ldots, y_K$ there is an almost weakly stable operator T with $\|Uy_k - Ty_k\| < \frac{\varepsilon}{4KM}$ and $\|U^*y_k - T^*y_k\| < \frac{\varepsilon}{4KM}$ for $M := \max_{k=1,\ldots,K, j=1,\ldots,n} |a_{jk}|$ and every $k = 1, \ldots, K$. Therefore we obtain

$$\|Ux_j - Tx_j\| \leq \sum_{k=1}^K |a_{jk}| \|Uy_k - Ty_k\| + 2\|o_j\| < \varepsilon$$

for every $j = 1, \ldots, n$. Analogously, $\|U^*x_j - T^*x_j\| < \varepsilon$ holds for every $j = 1, \ldots, n$, and the proposition is proved. $\qquad\square$

We can now prove the following category theorem for weakly and almost weakly stable unitary operators. To do so we extend the argument used by Halmos and Rohlin (see, e.g., [118, pp. 77–80]) for measure preserving transformations in ergodic theory.

Theorem 5.4. *The set $\mathcal{S}_\mathcal{U}$ of weakly stable unitary operators is of first category and the set $\mathcal{W}_\mathcal{U}$ of almost weakly stable unitary operators is residual in $\mathcal{U}$.*

Recall that a subset of a Baire space is called residual if its complement is of first category.

Proof. We first prove that $\mathcal{S}$ is of first category in $\mathcal{U}$. Fix $x \in H$ with $\|x\| = 1$ and consider the closed sets

$$M_k := \left\{ U \in \mathcal{U} : \; |\langle U^k x, x \rangle| \le \frac{1}{2} \right\}.$$

Let $U \in \mathcal{U}$ be weakly stable. Then there exists $n \in \mathbb{N}$ such that $U \in M_k$ for all $k \ge n$, i.e., $U \in \cap_{k \ge n} M_k$. So we obtain

$$\mathcal{S}_\mathcal{U} \subset \bigcup_{n=1}^{\infty} N_n, \tag{II.26}$$

where $N_n := \cap_{k \ge n} M_k$. Since the sets N_n are closed, it remains to show that $\mathcal{U} \setminus N_n$ is dense for every n.

Fix $n \in \mathbb{N}$ and let U be a periodic unitary operator. Then $U \notin M_k$ for some $k \ge n$ and therefore $U \notin N_n$. Since by Proposition 5.1 the periodic unitary operators are dense in $\mathcal{U}$, $\mathcal{S}$ is of first category.

To show that $\mathcal{W}_\mathcal{U}$ is residual we take a dense subset $D = \{x_j\}_{j=1}^{\infty}$ of H and define the open sets

$$W_{jkn} := \left\{ U \in \mathcal{U} : \; |\langle U^n x_j, x_j \rangle| < \frac{1}{k} \right\}.$$

Then the sets $W_{jk} := \cup_{n=1}^{\infty} W_{jkn}$ are also open.

We show that

$$\mathcal{W}_\mathcal{U} = \bigcap_{j,k=1}^{\infty} W_{jk} \tag{II.27}$$

holds.

The inclusion "$\subset$" follows from the definition of almost weak stability. To prove the converse inclusion we take $U \notin \mathcal{W}_\mathcal{U}$. Then there exists $x \in H$ with $\|x\| = 1$ and $\varphi \in \mathbb{R}$ such that $Ux = e^{i\varphi} x$. Take now $x_j \in D$ with $\|x_j - x\| \le \frac{1}{4}$. Then

$$\begin{aligned}
|\langle U^n x_j, x_j \rangle| &= |\langle U^n (x - x_j), x - x_j \rangle + \langle U^n x, x \rangle - \langle U^n x, x - x_j \rangle \\
&\quad - \langle U^n (x - x_j), x \rangle | \\
&\ge 1 - \|x - x_j\|^2 - 2\|x - x_j\| > \frac{1}{3}
\end{aligned}$$

holds for every $n \in \mathbb{N}$. So $U \notin W_{j3}$ which implies $U \notin \cap_{j,k=1}^{\infty} W_{jk}$, and therefore (II.27) holds. Moreover, all W_{jk} are dense by Proposition 5.3. Hence $\mathcal{W}_\mathcal{U}$ is residual as a countable intersection of open dense sets. $\square$

5.2 Isometries

We now consider the space $\mathcal{I}$ of all isometries on H, a separable and infinite-dimensional Hilbert space, endowed with the strong operator topology and prove analogous category results as above. Note that $\mathcal{I}$ is a complete metric space with respect to the metric given by the formula

$$d(T,S) := \sum_{j=1}^{\infty} \frac{\|Tx_j - Sx_j\|}{2^j \|x_j\|} \quad \text{for } T, S \in \mathcal{I},$$

where $\{x_j\}_{j=1}^{\infty}$ is a fixed dense subset of $H \setminus \{0\}$.

Further we denote by $\mathcal{S}_{\mathcal{I}}$ the set of all weakly stable isometries on H and by $\mathcal{W}_{\mathcal{I}}$ the set all almost weakly stable isometries on H.

The results in this subsection are based on the following classical theorem on isometries on Hilbert spaces, see [238, Theorem 1.1].

Theorem 5.5 (Wold decomposition). *Let V be an isometry on a Hilbert space H. Then H can be decomposed into an orthogonal sum $H = H_0 \oplus H_1$ of V-invariant subspaces such that the restriction of V on H_0 is unitary and the restriction of V on H_1 is a unilateral shift, i.e., there exists a subspace $Y \subset H_1$ with $V^n Y \perp V^m Y$ for all $n \neq m$, $n, m \in \mathbb{N}$, such that $H_1 = \oplus_{n=1}^{\infty} V^n Y$ holds.*

We need the following easy lemma, see also Peller [211].

Lemma 5.6. *Let Y be a Hilbert space and let R be the right shift on $H := l^2(\mathbb{N}, Y)$. Then there exists a sequence $\{T_n\}_{n=1}^{\infty}$ of periodic unitary operators on H converging strongly to R.*

Proof. We define the operators T_n by

$$T_n(x_1, x_2, \ldots, x_n, \ldots) := (x_n, x_1, x_2, \ldots, x_{n-1}, x_{n+1}, \ldots).$$

Every T_n is unitary and has period n. Moreover, for an arbitrary $x = (x_1, x_2, \ldots) \in H$ we have

$$\|T_n x - Rx\|^2 = \|x_n\|^2 + \sum_{k=n}^{\infty} \|x_{k+1} - x_k\|^2 \xrightarrow[n \to \infty]{} 0,$$

and the lemma is proved. $\square$

As a first application of the Wold decomposition we obtain the density of the periodic operators in $\mathcal{I}$. (Note that periodic isometries are unitary.)

Proposition 5.7. *The set of all periodic isometries is dense in $\mathcal{I}$.*

Proof. Let V be an isometry on H. Then by Theorem 5.5 the orthogonal decomposition $H = H_0 \oplus H_1$ holds, where the restriction V_0 on H_0 is unitary and the space H_1 is unitarily equivalent to $l^2(\mathbb{N}, Y)$. The restriction V_1 of V on H_1 corresponds (by this equivalence) to the right shift operator on $l^2(\mathbb{N}, Y)$. By Proposition 5.1 and Lemma 5.6 we can approximate both operators V_0 and V_1 by unitary periodic ones and the assertion follows. $\square$

We further obtain the density of the set of all almost weakly stable operators in $\mathcal{I}$.

Proposition 5.8. *The set $\mathcal{W}_\mathcal{I}$ of almost weakly stable isometries is dense in $\mathcal{I}$.*

Proof. Let V be an isometry on H and let V_0 and V_1 be the corresponding restrictions of V to the orthogonal subspaces H_0 and H_1 from Theorem 5.5. By Lemma 5.6 the operator V_1 can be approximated by unitary operators on H_1. The assertion now follows from Proposition 5.3. □

Using the same idea as in the proof of Theorem 5.4 one obtains, using Propositions 5.7 and 5.8, the following category result for weakly and almost weakly stable isometries.

Theorem 5.9. *The set $\mathcal{S}_\mathcal{I}$ of all weakly stable isometries is of first category and the set $\mathcal{W}_\mathcal{I}$ of all almost weakly stable isometries is residual in $\mathcal{I}$.*

5.3 Contractions

The above category results are now extended to the case of contractive operators.

Let the Hilbert space H be as before and denote by $\mathcal{C}$ the set of all contractions on H endowed with the weak operator topology. Note that $\mathcal{C}$ is a complete metric space with respect to the metric given by the formula

$$d(T, S) := \sum_{i,j=1}^{\infty} \frac{|\langle Tx_i, x_j \rangle - \langle Sx_i, x_j \rangle|}{2^{i+j} \|x_i\| \|x_j\|} \quad \text{for } T, S \in \mathcal{C},$$

where $\{x_j\}_{j=1}^{\infty}$ is a fixed dense subset of H with each $x_j \neq 0$.

By Takesaki [239, p. 99], the set of all unitary operators is dense in $\mathcal{C}$ (see also Peller [211] for a much stronger assertion). Combining this with Propositions 5.1 and 5.3 we have the following fact.

Proposition 5.10. *The set of all periodic unitary operators and the set of all almost weakly stable unitary operators are both dense in $\mathcal{C}$.*

The following elementary property is a key for the further results (cf. Halmos [121, p. 14]).

Lemma 5.11. *Let $\{T_n\}_{n=1}^{\infty}$ be a sequence of linear operators on a Hilbert space H converging weakly to a linear operator S. If $\|T_n x\| \leq \|Sx\|$ for every $x \in H$, then $\lim_{n \to \infty} T_n = S$ strongly.*

Proof. For each $x \in H$ we have

$$\|T_n x - Sx\|^2 = \langle T_n x - Sx, T_n x - Sx \rangle = \|Sx\|^2 + \|T_n x\|^2 - 2\mathrm{Re}\,\langle T_n x, Sx \rangle$$
$$\leq 2\langle Sx, Sx \rangle - 2\mathrm{Re}\,\langle T_n x, Sx \rangle = 2\mathrm{Re}\,\langle (S - T_n)x, Sx \rangle \xrightarrow[n \to \infty]{} 0,$$

and the lemma is proved. □

We now state the category result for contractions, but note that its proof differs from the corresponding proofs in the two previous subsections.

Theorem 5.12. *The set S_C of all weakly stable contractions is of first category and the set W_C of all almost weakly stable contractions is residual in C.*

Proof. To prove the first statement we fix $x \in X$, $\|x\| = 1$, and define as before the sets

$$N_n := \left\{ T \in C : |\langle T^k x, x \rangle| \leq \frac{1}{2} \text{ for all } k \geq n \right\}.$$

Let $T \in C$ be weakly stable. Then there exists $n \in \mathbb{N}$ such that $T \in N_n$, and we obtain

$$S_C \subset \bigcup_{n=1}^{\infty} N_n. \tag{II.28}$$

It remains to show that the sets N_n are nowhere dense. Fix $n \in \mathbb{N}$ and let U be a periodic unitary operator. We show that U does not belong to the closure of N_n. Assume the opposite, i.e., that there exists a sequence $\{T_k\}_{k \in \mathbb{N}} \subset N_n$ satisfying $\lim_{k \to \infty} T_k = U$ weakly. Then, by Lemma 5.11, $\lim_{k \to \infty} T_k = U$ strongly and therefore $U \in N_n$ by the definition of N_n. This contradicts the periodicity of U. By the density of the set of unitary periodic operators in C we obtain that N_n is nowhere dense and therefore S_C is of first category.

To show the residuality of W_C we again take a dense subset $D = \{x_j\}_{j=1}^{\infty}$ of H and define

$$W_{jk} := \left\{ T \in C : |\langle T^n x_j, x_j \rangle| < \frac{1}{k} \text{ for some } n \in \mathbb{N} \right\}.$$

As in the proof of Theorem 5.4 the equality

$$W_C = \bigcap_{j,k=1}^{\infty} W_{jk} \tag{II.29}$$

holds.

Fix $j, k \in \mathbb{N}$. We have to show that the complement W_{jk}^c of W_{jk} which coincides with

$$\left\{ T \in C : |\langle T^n x_j, x_j \rangle| \geq \frac{1}{k} \text{ for all } n \in \mathbb{N} \right\}$$

is nowhere dense. Let U be a unitary almost weakly stable operator. Assume that there exists a sequence $\{T_m\}_{m=1}^{\infty} \subset W_{jk}^c$ satisfying $\lim_{m \to \infty} T_m = U$ weakly. Then, by Lemma 5.11, $\lim_{m \to \infty} T_m = U$ strongly and therefore $U \in W_{jk}^c$. This contradicts the almost weak stability of U. Therefore the set of all unitary almost weakly stable operators does not intersect the closure of W_{jk}^c. By Proposition 5.10 all the sets W_{jk}^c are nowhere dense and therefore W_C is residual. $\square$

Remark 5.13. As a consequence of Theorem 5.12 we see that the set of all strongly stable operators as well as the set of all operators T satisfying $r(T) < 1$ are also of first category in $\mathcal{C}$.

Open question 5.14. Do the above category theorems hold on reflexive Banach spaces?

On non-reflexive Banach spaces these results are not true in general. Indeed, every almost weakly stable contraction on the space l^1 is automatically weakly and even strongly stable by Schur's lemma, see, e.g., Conway [52, Prop. V.5.2], and Lemma 2.4.

For related category results see e.g. Iwanik [82], Bartoszek [16], Bartoszek, Kuna [18].

Remark 5.15. In Section IV.3 we prove a stronger version of the above results.

6   Stability via Lyapunov's equation

Our spectral and resolvent criteria for stability and power boundedness of an operator T used information on $R(\lambda, T)$ in many points $\lambda \in \rho(T)$, i.e., one had to solve the equation

$$(\lambda I - T)x = y$$

for all y and infinitely many λ (and then even had to estimate the solutions). This is, in practical situations, a quite difficult task.

In a Hilbert space H, one can reduce this problem to the solution of a single equation in a new, but higher dimensional space. The idea is to use the implemented operators on and the order structure of $\mathcal{L}(H)$, see Section I.4. Recall that implemented operators are positive operators on $\mathcal{L}(H)$ defined as $\mathcal{T}S := T^*ST$ for some fixed operator $T \in \mathcal{L}(H)$.

6.1   Uniform exponential and strong stability

We start by characterising uniform exponential stability.

Theorem 6.1. *Let T be a bounded operator on a Hilbert space H and $\mathcal{T}$ its implemented operator on $\mathcal{L}(H)$. Then the following assertions are equivalent.*

(i) *T is uniformly exponentially stable on H.*

(ii) *$\mathcal{T}$ is uniformly exponentially stable on $\mathcal{L}(H)$.*

(iii) *$1 \in \rho(\mathcal{T})$ and $0 \leq R(1, \mathcal{T})$.*

(iv) *There exists $0 \leq Q \in \mathcal{L}(H)$ such that*

$$(\mathcal{T} - I)Q = T^*QT - Q = -I. \tag{II.30}$$

(v) *There exists $0 \leq Q \in \mathcal{L}(H)$ such that $T^*QT - Q \leq -I$.*

In this case, Q in (iv) *is unique and given by $Q = R(1,T)I = \sum_{n=0}^{\infty} T^{*n}T^n$.*

Equation (II.30) is called *Lyapunov equation.*

Proof. (i)$\Leftrightarrow$(ii) follows from Lemma I.4.6, (ii)$\Leftrightarrow$(iii) by Theorem I.4.4 and (iii) implies (iv) by taking $Q := R(1,T)I \geq 0$. Since (iv)$\Rightarrow$(v) is trivial, we now assume (v) and prove (iii). Take $0 \leq S$ on H and observe that $S \leq \|S\| \cdot I$. Since $(I - T)Q = I$ we obtain

$$0 \leq \sum_{k=0}^{n} T^k S \leq \|S\| \sum_{k=0}^{n} T^k I = \|S\| \sum_{k=0}^{n} T^k (I - T)Q = \|S\|(Q - T^{n+1}Q) \leq \|S\|Q.$$

This shows that the sequence $\{\sum_{k=0}^{n} T^k S\}_{n=0}^{\infty}$ is monotonically increasing and bounded, hence the series $\sum_{k=0}^{\infty} T^k S$ converges weakly for every $0 \leq S$ and hence for every $S \in \mathcal{L}(H)$ by Lemma I.4.1. So $1 \in \rho(T)$ and $R(1,T) = \sum_{k=0}^{\infty} T^k$ (for the strong operator topology) implying $R(1,T) \geq 0$, and (c) is proved. $\square$

We now turn our attention to strong stability.

Theorem 6.2. *Let $T \in \mathcal{L}(H)$ and $T \in \mathcal{L}(\mathcal{L}(H))$ be as above. Then the following assertions are equivalent.*

(i) *T is strongly stable.*

(ii) *$\lim_{n\to\infty}\langle T^n Sx, x\rangle = 0$ for every $S \in \mathcal{L}(H)$ and $x \in H$.*

(iii) *For every $a > 0$ there exist (unique) $0 \leq Q_a, \tilde{Q}_a \in \mathcal{L}(H)$ satisfying*

$$e^{-2a}T^*Q_a T - Q_a = -I \quad and \quad e^{-2a}T\tilde{Q}_a T^* - \tilde{Q}_a = -I$$

such that

$$\lim_{a\to 0+} a\langle Q_a x, x\rangle = 0 \quad and \quad \sup_{a>0} a\langle \tilde{Q}_a x, x\rangle < \infty \quad for\ all\ x \in H. \quad (II.31)$$

The operators Q_a and $\tilde{Q}_a$ are obtained as

$$Q_a = e^{2a} R(e^{2a}, T)I \quad and \quad \tilde{Q}_a = e^{2a} R(e^{2a}, T^*)I.$$

Here, T^* denotes the operator implemented by T^*.

Proof. The implication (ii)$\Rightarrow$(i) follows from the relation $\|Tx\|^2 = \langle TIx, x\rangle$. To prove the converse take $S \in \mathcal{L}(H)_{sa}$ and observe $-cI \leq S \leq cI$ for $c := \|S\|$. This implies $|\langle T^n Sx, x\rangle| \leq c|\langle T^n Ix, x\rangle| = c\|T^n x\|^2$ proving (i)$\Rightarrow$(ii).

Assume that (i) or (iii) hold. Then Theorem 6.1 applied to $e^{-a}T$ and its implemented operator $T_a = e^{-2a}T$ implies that $Q_a := R(1, T_a)I = e^{2a} R(e^{2a}, T)I$

is the unique solution of the first rescaled Lyapunov equation in (iii) for every $a > 0$. Moreover,

$$\langle Q_a x, x \rangle = \sum_{n=0}^{\infty} e^{-2an} \langle T^n I x, x \rangle = \sum_{n=0}^{\infty} e^{-2an} \| T^n x \|^2.$$

Analogously, $\tilde{Q}_a := e^{2a} R(e^{2a}, T^*) I$ is the unique solution of the second rescaled Lyapunov equation in (iii) and satisfies $\langle \tilde{Q}_a x, x \rangle = \sum_{n=0}^{\infty} e^{-2an} \| T^n x \|^2$.

We now see that (II.31) is equivalent to the property

$$\lim_{a \to 0+} a \sum_{n=0}^{\infty} e^{-2an} \| T^n x \|^2 = 0 \quad \text{and} \quad \sup_{a>0} a \sum_{n=0}^{\infty} e^{-2an} \| T^{*n} x \|^2 < \infty.$$

By Lemma I.2.6, this is equivalent to

$$\lim_{n \to \infty} \frac{1}{n+1} \sum_{k=0}^{n} \| T^k x \|^2 = 0 \quad \text{and} \quad \sup_{n>0} \frac{1}{n+1} \sum_{k=0}^{n} \| T^{*k} x \|^2 < \infty$$

for every $x \in H$ which is equivalent to strong stability of T by Proposition 2.7, and therefore (i)$\Leftrightarrow$(iii). $\square$

Remark 6.3. Note that one can again replace "$= -I$" by "$\leq -I$" in the rescaled Lyapunov equations in (iii).

6.2   Power boundedness

One can also characterise power boundedness of operators on Hilbert spaces via analogous Lyapunov equations.

Theorem 6.4. *For $T \in \mathcal{L}(H)$ and $\mathcal{T} \in \mathcal{L}(\mathcal{L}(H))$ as before, the following assertions are equivalent.*

(i) *T is power bounded.*

(ii) *$\mathcal{T}$ is power bounded.*

(iii) *For every $a > 0$ there exist (unique) $0 \leq Q_a, \tilde{Q}_a \in \mathcal{L}(H)$ satisfying*

$$e^{-2a} T^* Q_a T - Q_a = -I \quad \text{and} \quad e^{-2a} T \tilde{Q}_a T^* - Q_a = -I$$

such that

$$\sup_{a>0} a \langle Q_a x, x \rangle < \infty \quad \text{and} \quad \sup_{a>0} a \langle \tilde{Q}_a x, x \rangle < \infty \quad \text{for all } x \in H. \quad \text{(II.32)}$$

In this case, one has $Q_a = e^{2a} R(e^{2a}, T) I$ and $\tilde{Q}_a = e^{2a} R(e^{2a}, T^*) I$.

The implication (i)$\Leftrightarrow$(ii) follows directly from Lemma I.4.6, while the proof of (i)$\Leftrightarrow$(iii) is analogous to the proof of Theorem 6.2 using Proposition 1.7.

Final remark

Again, weak stability is much more delicate to treat than strong stability or power boundedness. The main problem occurs when T is a unitary operator. (Note that by Theorem 3.9, the unitary part of a contraction is decisive for weak stability.) In this case, the implemented operator $\mathcal{T}$ satisfies $\mathcal{T}I = I$, and therefore weak stability cannot be described in terms of $\{\mathcal{T}^n I\}_{n=0}^{\infty}$ or $R(\cdot, \mathcal{T})I$, respectively, as we could do in Theorems 6.1 and 6.2.

Open question 6.5. Is it possible to characterise weak stability of operators on Hilbert spaces via some kind of Lyapunov equation?

Chapter III

Stability of C_0-semigroups

In this chapter we discuss boundedness, polynomial boundedness, exponential, strong, weak and almost weak stability of C_0-semigroups $(T(t))_{t\geq 0}$ on Banach spaces.

The goal is to characterise these properties without using the semigroup explicitly. The optimal case is to have a characterisation in terms of the spectrum of the generator A. However, such results are possible only in very special cases. As it turns out, a good substitute is the behaviour of the resolvent $R(\lambda, A)$ in a neighbourhood of $i\mathbb{R}$.

In our presentation we emphasise the similarities and differences to the discrete case treated in Chapter II.

1 Boundedness

In this section we consider boundedness and the related notion of polynomial boundedness for C_0-semigroups. While boundedness of C_0-semigroups is difficult to check if the semigroup is not contractive, the weaker notion of polynomial boundedness is easier to characterise.

1.1 Preliminaries

We start with bounded C_0-semigroups and their elementary properties.

Definition 1.1. A C_0-semigroup $T(\cdot)$ on a Banach space X is called *bounded* if $\sup_{t\geq 0} \|T(t)\| < \infty$.

Remark 1.2. Every bounded semigroup $T(\cdot)$ satisfies $\omega_0(T) \leq 0$ and hence $\sigma(A) \subset \{z : \operatorname{Re}(z) \leq 0\}$. However, the spectral condition $\sigma(A) \subset \{z : \operatorname{Re}(z) \leq 0\}$ does not imply boundedness of the semigroup, as can be seen already from the matrix semigroup $T(\cdot)$ given by $T(t) = \begin{pmatrix} 1 & t \\ 0 & 1 \end{pmatrix}$ on $\mathbb{C}^2$. In Subsection 1.3 we give more sophisticated examples and a general description of possible growth.

Remark 1.3. It is interesting that, analogous to power bounded operators (see Subsection 1.1), more information on the spectrum of the generator of a bounded C_0-semigroup is known if X is separable. For example, Jamison [135] proved that if a C_0-semigroup on a separable Banach space is bounded, then the point spectrum of its generator on the imaginary axis has to be at most countable. For more results in this direction see Ransford [218], Ransford, Roginskaya [219] and Badea, Grivaux [14].

   We will see later that countability of the spectrum of the generator on the imaginary axis plays an important role for strong stability of the semigroup, see Subsection 3.3.

   The following simple lemma is useful to understand boundedness.

Lemma 1.4. Let $T(\cdot)$ be a bounded semigroup on a Banach space X. Then there exists an equivalent norm on X such that $T(\cdot)$ becomes a contraction semigroup.

Proof. Take $\|x\|_1 := \sup_{t \geq 0} \|T(t)x\|$ for every $x \in X$. $\square$

Remark 1.5. Packel [207] showed that not all bounded C_0-semigroups on Hilbert spaces are similar to a contraction semigroup for a Hilbert space norm. His example was a modification of the corresponding examples of Foguel [87] and Halmos [120] in the discrete case.

   However, Sz.-Nagy [236] proved that every bounded C_0-group is similar to a unitary one. We also mention Vũ and Yao [253] who proved that every bounded uniformly continuous quasi-compact C_0-semigroup on a Hilbert space is similar to a contraction semigroup.

   Zwart [263] and Guo, Zwart [113, Thm. 8.2] showed that some kind of Cesàro boundedness implies boundedness. See also van Casteren [45] for the case of bounded C_0-groups on Hilbert spaces.

Theorem 1.6. For a C_0-semigroup $T(\cdot)$ on a Banach space X the following assertions are equivalent.

(a) $T(\cdot)$ is bounded.

(b) For all $x \in X$ and $y \in X'$ we have

$$\sup_{t \geq 0} \frac{1}{t} \int_0^t \|T(s)x\|^2 \, ds < \infty,$$

$$\sup_{t \geq 0} \frac{1}{t} \int_0^t \|T'(s)y\|^2 \, ds < \infty.$$

Proof. (a)$\Rightarrow$(b) is clear.

(b)$\Rightarrow$(a). Take $x \in X$ and $y \in X'$. Then we have by the Cauchy–Schwarz inequality

$$|\langle T(t)x, y \rangle| = \frac{1}{t} \int_0^t |\langle T(t)x, y \rangle| \, ds = \frac{1}{t} \int_0^t |\langle T(t-s)x, T'(s)y \rangle| \, ds$$

$$\leq \left(\frac{1}{t} \int_0^t \|T(s)x\|^2 \, ds \right)^{\frac{1}{2}} \left(\frac{1}{t} \int_0^t \|T'(s)y\|^2 \, ds \right)^{\frac{1}{2}},$$

and hence every weak orbit $\{\langle T(t)x, y \rangle : t \geq 0\}$ is bounded by assumption. So $T(\cdot)$ is bounded by the uniform boundedness principle. □

Remark 1.7. The second part of condition (b) in the above theorem cannot be omitted, i.e., absolute Cesàro-boundedness is not equivalent to boundedness. Van Casteren [45] gave an example of an unbounded C_0-group on a Hilbert space satisfying the first part of condition (b).

Theorem 1.6 can also be formulated for single orbits.

Proposition 1.8. *For a C_0-semigroup $T(\cdot)$ on a Banach space X, $x \in X$ and $y \in X'$ the weak orbit $\{\langle T(t)x, y \rangle : t \geq 0\}$ is bounded if and only if there exist $p, q > 1$ with $\frac{1}{p} + \frac{1}{q} = 1$ such that*

$$\sup_{t \geq 0} \frac{1}{t} \int_0^t \|T(s)x\|^p \, ds < \infty,$$

$$\sup_{t \geq 0} \frac{1}{t} \int_0^t \|T'(s)y\|^q \, ds < \infty.$$

However, since the semigroup is, in most cases, not known explicitly, it is important to find characterisations not involving it directly.

1.2 Representation as shift semigroups

It is well-known that every bounded semigroup on a Banach space is similar to the left shift semigroup on a space of continuous functions.

Theorem 1.9. *Let $T(\cdot)$ be a contractive C_0-semigroup on a Banach space X. If $T(\cdot)$ is strongly stable, then it is isometrically equivalent to the left shift semigroup on a closed subspace of $BUC(\mathbb{R}_+, X)$.*

By $BUC(\mathbb{R}_+, X)$ we denote the space of all bounded, uniformly continuous functions from $\mathbb{R}_+$ to X.

Proof. Define the operator $J : X \to BUC(\mathbb{R}_+, X)$ by

$$(Jx)(s) := T(s)x, \quad s \geq 0, \ x \in X.$$

This operator identifies a vector x with its orbit under the semigroup. (Note that every function $s \mapsto T(s)x$ is uniformly continuous by the assumption on $T(\cdot)$.) By contractivity of $T(\cdot)$ we have $\|x\| = \max_{s \geq 0} \|T(s)x\| = \|Jx\|$ and hence J is an isometry. Moreover,

$$JT(t)x = T(t + \cdot)x, \quad t \geq 0, \ x \in X,$$

i.e., $T(\cdot)$ corresponds to the left shift semigroup on $\operatorname{rg} J$ which is a closed shift-invariant subspace of $C_0(\mathbb{R}_+, X)$.   □

By the renorming procedure from Lemma 1.4 we obtain the following general form of bounded semigroups.

Corollary 1.10. *Let $T(\cdot)$ be a strongly stable semigroup on a Banach space X. Then $T(\cdot)$ is isomorphic to the left shift semigroup on a closed subspace of $BUC(\mathbb{R}_+, X_1)$, where $X_1 = X$ endowed with an equivalent norm.*

1.3   Characterisation via resolvent

In this subsection we characterise boundedness for C_0-semigroups by the first or second power of the resolvent of its generator.

Theorem 1.11 (Gomilko [104], Shi and Feng [231]). *Let A be a densely defined operator on a Banach space X satisfying $s_0(A) \leq 0$. Consider the following assertions.*

(a) *For every $x \in X$ and $y \in X'$,*

$$\sup_{a > 0} a \int_{-\infty}^{\infty} \|R(a + is, A)x\|^2 \, ds < \infty,$$

$$\sup_{a > 0} a \int_{-\infty}^{\infty} \|R(a + is, A')y\|^2 \, ds < \infty;$$

(b) $\displaystyle \sup_{a > 0} a \int_{-\infty}^{\infty} |\langle R^2(a + is, A)x, y \rangle| \, ds < \infty$ *for all $x \in X$, $y \in X'$;*

(c) *A generates a bounded C_0-semigroup on X.*

Then (a)$\Rightarrow$(b)$\Rightarrow$(c). Moreover, if X is a Hilbert space, then (a)$\Leftrightarrow$(b)$\Leftrightarrow$(c).

Proof. Assume (a). By the Cauchy-Schwarz inequality we have

$$a \int_{-\infty}^{\infty} |\langle R^2(a + is, A)x, y \rangle| \, ds \leq a \int_{-\infty}^{\infty} \|R(a + is, A)x\| \|R(a + is, A')y\| \, ds$$

$$\leq \left(a \int_{-\infty}^{\infty} \|R(a + is, A)x\|^2 \, ds \right)^{\frac{1}{2}} \left(a \int_{-\infty}^{\infty} \|R(a + is, A)x\|^2 \, ds \right)^{\frac{1}{2}}$$

which proves (b).

Consider now the implication (b)⇒(c). We will construct the semigroup explicitly using the idea from Shi, Feng [231].

We first prove that (b) implies $s_0(A) \leq 0$. Since $\frac{d}{dz}R(z,A) = -R^2(z,A)$, we have for all $a > 0$, $x \in X$ and $y \in X'$ that

$$\langle R(a+is, A)x, y \rangle = \langle R(a, A)x, y \rangle - i \int_0^s \langle R(a+i\tau, A)^2 x, y \rangle \, d\tau. \tag{III.1}$$

By the absolute convergence of the integral on the right-hand side we obtain that $\lim_{s \to \infty} \langle R(a+is, A)x, y \rangle = 0$. From (III.1) and (b) it follows that

$$\|R(a+is, A)\| \leq \frac{M}{a},$$

hence $s_0(A) \leq 0$ holds.

Define now $T(0) = I$ and

$$T(t)x = \frac{1}{2\pi t} \int_{-\infty}^{\infty} e^{(a+is)t} R(a+is, A)^2 x \, ds$$

for all $x \in X$ and $t > 0$. By the assumption the operators $T(\cdot)$ are well-defined and form a semigroup by Theorem I.3.9. Let us estimate $\|T(t)\|$. By (b) and the uniform boundedness principle we have

$$|\langle T(t)x, y \rangle| \leq \frac{e^{at}}{2\pi t} \int_{-\infty}^{\infty} |\langle R(a+is, A)^2 x, y \rangle| \, ds$$

$$\leq \frac{M e^{at}}{2\pi t a} \|x\| \|y\|.$$

Taking $a := t^{-1}$ we obtain for $K := \frac{Me}{2\pi}$ the desired estimate

$$\|T(t)\| \leq K \quad \forall t \geq 0. \tag{III.2}$$

Finally, this and Theorem I.3.9 imply the strong continuity of $T(\cdot)$.

For the implication (c)⇒(a), let $T(\cdot)$ be a C_0-semigroup on a Hilbert space H with generator A satisfying $K := \sup_{t \geq 0} \|T(t)\| < \infty$. By Parseval's inequality and integration by parts we have

$$\int_{-\infty}^{\infty} \|R(a+is, A)x\|^2 \, ds = \int_0^{\infty} e^{-2at} \|T(t)x\|^2 \, dt \leq K^2 \int_0^{\infty} e^{-2at} \|x\|^2 \, dt \leq \frac{K^2}{2a} \|x\|^2$$

and analogously

$$\int_{-\infty}^{\infty} \|R(a+is, A')y\|^2 \, ds \leq \frac{K^2}{2a} \|y\|^2$$

for every $x, y \in H$, and the theorem is proved. □

Remarks 1.12. 1) Gomilko first proved the non-trivial implication (b)$\Rightarrow$(c) using the Hille–Yosida theorem. Then Shi and Feng gave an alternative proof using an explicit construction of the semigroup by formula (I.13) in Subsection I.3.2. The direction (c)$\Rightarrow$(a) for Hilbert spaces follows easily from Plancherel's theorem.

Note that van Casteren [45] presented an analogous characterisation for bounded C_0-groups on Hilbert spaces much earlier. He also showed that the second part of condition (a) cannot be omitted.

2) One can replace the L^2-norms in condition (a) by the L^p-norm in the first inequality and L^q-norm in the second one for $p, q > 1$ with $\frac{1}{p} + \frac{1}{q} = 1$, possibly depending on x and y. However, for the converse implication, if X is a Hilbert space, one needs $p = q = 2$.

A direct consequence of the construction given in the proof of Theorem 1.11 is the following result for single orbits.

Proposition 1.13. *Let X be Banach space and A generate a C_0-semigroup $T(\cdot)$ with $s_0(A) \leq 0$, $x \in X$ and $y \in X'$. Consider the following assertions.*

(a) *For some $p, q > 1$ with $\frac{1}{p} + \frac{1}{q} = 1$,*

$$\limsup_{a \to 0+} a \int_{-\infty}^{\infty} \|R(a + is, A)x\|^p \, ds < \infty,$$

$$\limsup_{a \to 0+} a \int_{-\infty}^{\infty} \|R(a + is, A')y\|^q \, ds < \infty;$$

(b) $\displaystyle \limsup_{a \to 0+} a \int_{-\infty}^{\infty} |\langle R^2(a + is, A)x, y \rangle| \, ds < \infty;$

(c) $\{\langle T(t)x, y \rangle : t \geq 0\}$ *is bounded.*

Then (a)$\Rightarrow$(b)$\Rightarrow$(c). *Moreover, if X is a Hilbert space, then* (a)$\Leftrightarrow$(b)$\Leftrightarrow$(c) *for $p = q = 2$.*

In particular, conditions (a) *and* (b) *holding for all $x \in X$ and $y \in X'$ both imply boundedness of $T(\cdot)$ and are equivalent to it when X is a Hilbert space and $p = q = 2$.*

Note that conditions (a) and (b) in Theorem 1.11 are not necessary for boundedness since Corollary 2.19 below shows that the integrals in (a) and (b) can diverge in general. But even if all these integrals converge, condition (a) is still not necessary for boundedness even for isometric groups with bounded generators. This can be seen from the following example.

Example 1.14. On $X := C[0,1]$ consider the bounded operator A defined by $Af(s) := isf(s)$ which generates the isometric (and hence bounded) group given by $T(t)f(s) = e^{ist}f(s)$. Then for every $a > 0$ and $b \in [0,1]$ we have

$$\|R(a + ib, A)\mathbf{1}\| = \sup_{s \in [0,1]} \frac{1}{|a + ib - is|} = \frac{1}{a}$$

which implies

$$a \int_{-\infty}^{\infty} \|R(a+ib,A)\mathbf{1}\|^2 \, db \geq a \int_0^1 \|R(a+ib,A)\mathbf{1}\|^2 \, db = \frac{1}{a} \to \infty \quad \text{as } a \to 0+ \, .$$

So A does not satisfy condition (a) in Theorem 1.11 as well as condition (a) in Proposition 1.13.

Since the same example works on the space $L^\infty[0,1]$, we can take the pre-adjoint and see that also on the space $L^1([0,1])$ the operator A does not satisfy (a) in Theorem 1.11.

It is still open to find analogous necessary and sufficient resolvent conditions for C_0-semigroups on Banach spaces.

1.4 Polynomial boundedness

In this subsection we introduce and discuss polynomial boundedness of C_0-semigroups. Surprisingly, this weaker notion allows a much simpler characterisation.

Definition 1.15. A semigroup $T(\cdot)$ on a Banach space X is called *polynomially bounded* if $\|T(t)\| \leq p(t)$, $t \geq 0$, for some polynomial p.

In the following we will assume $\|T(t)\| \leq K(1+t^d)$ for some constants $d \geq 0$ (being not necessarily an integer) and $K \geq 1$.

Note that every polynomially bounded semigroup $T(\cdot)$ still satisfies $\omega_0(T) \leq 0$. The following example shows that the converse implication does not hold.

Example 1.16 (C_0-**semigroups satisfying** $\omega_0(T) \leq 0$ **with non-polynomial growth**). The following construction is analogous to Example II.1.16 in the discrete case.

Consider the Hilbert space

$$H := L_{a^2}^2 = \{f : \mathbb{R}_+ \to \mathbb{C} \text{ measurable} : \int_0^\infty |f(s)|^2 a^2(s) \, ds < \infty\}$$

for some positive continuous function a satisfying $a(0) \geq 1$ and

$$a(t+s) \leq a(t)a(s) \qquad \text{for all } t,s \in \mathbb{R}_+ \tag{III.3}$$

with the corresponding natural scalar product. On H take the right shift semigroup $T(\cdot)$.

We first check strong continuity of $T(\cdot)$. For the characteristic function f on an interval $[a,b]$ and $t < b - a$ one has

$$\|T(t)f - f\| = \int_a^{a+t} a^2(s) \, ds + \int_b^{b+t} a^2(s) \, ds \xrightarrow[t \to 0+]{} 0.$$

Further, for $f \in X$ we have by (III.3)

$$\|T(t)f\|^2 = \int_t^\infty |f(s-t)|^2 a^2(s)\, ds = \int_0^\infty |f(s)|^2 a^2(s+t)\, ds$$

$$\leq a^2(t) \int_0^\infty |f(s)|^2 a^2(s)\, ds = a^2(t) \|f\|^2$$

for every $t \geq 0$ and therefore $\|T(t)\| \leq a(t)$. A density argument shows that the semigroup $T(\cdot)$ is strongly continuous.

Moreover, for the characteristic functions f_n of the intervals $[0, 1/n]$ we have

$$\|T(t)f_n\|^2 = \int_t^{t+\frac{1}{n}} a^2(s)\, ds = \frac{\int_t^{t+\frac{1}{n}} a^2(s)\, ds}{\int_0^{\frac{1}{n}} a^2(s)\, ds} \|f_n\|^2$$

and hence

$$\|T(t)\|^2 \geq \frac{\frac{1}{n}\int_t^{t+\frac{1}{n}} a^2(s)\, ds}{\frac{1}{n}\int_0^{\frac{1}{n}} a^2(s)\, ds} \xrightarrow[n\to\infty]{} \frac{a^2(t)}{a^2(0)}.$$

So we obtain the norm estimate

$$\frac{a(t)}{a(0)} \leq \|T(t)\| \leq a(t)$$

and hence the semigroup $T(\cdot)$ has the same growth as the function a. Note that if $a(0) = 1$, then $\|T(t)\| = a(t)$ for all $t \geq 0$. Now every function a satisfying (III.3) and growing faster than every polynomial but slower than any exponential function with positive exponent yields an example of a non-polynomially growing C_0-semigroup $T(\cdot)$ satisfying $\omega_0(T) \leq 0$.

As a concrete example of such a function take

$$a(t) := (t+6)^{\ln(t+6)} = e^{\ln^2(t+6)}$$

(see Example II.1.16), from which we obtain a C_0-semigroup growing like $t^{\ln t}$. Analogously, one can construct a C_0-semigroup growing as $t^{\ln^\alpha t}$ for any $\alpha \geq 1$.

The idea to use condition (III.3) belongs to Sen-Zhong Huang (private communication).

The following theorem characterises operators generating polynomially bounded C_0-semigroups, see Eisner [64]. This generalises Theorem 1.11 for bounded C_0-semigroups and a theorem of Malejki [183] (see also Kiselev [150]) in the case of C_0-groups. The proof is a modification of the one of Theorem 1.11, see [64].

Theorem 1.17. *Let X be a Banach space and A be a densely defined operator on X with $s(A) \leq 0$ and $d \in [0, \infty)$. If the condition*

$$\int_{-\infty}^\infty |\langle R(a+is, A)^2 x, y\rangle|\, ds \leq \frac{M}{a}(1+a^{-d})\|x\|\|y\| \qquad \forall x \in X,\ \forall y \in X' \quad \text{(III.4)}$$

holds for all $a > 0$, then A is the generator of a C_0-semigroup $(T(t))_{t \geq 0}$ which does not grow faster than t^d, i.e.,

$$\|T(t)\| \leq K(1 + t^d) \tag{III.5}$$

for some constant K and all $t > 0$. Conversely, if X is a Hilbert space, then the growth condition (III.5) implies (III.4) for the parameter $d_1 := 2d$.

Remark 1.18. Kiselev [150] showed that the exponent $2d$ in the implication (III.5) $\Rightarrow$ (III.4) is sharp for the case of C_0-groups.

Remark 1.19. Analogously to the bounded case, conditions

$$\int_{-\infty}^{\infty} \|R(a + is, A)x\|^p \, ds \leq \frac{M}{a}(1 + a^{-d}) \quad \text{for every } x \in X,$$

$$\int_{-\infty}^{\infty} \|R(a + is, A')y\|^q \, ds \leq \frac{M}{a}(1 + a^{-d}) \quad \text{for every } y \in X'$$

for some $M \geq 1$ and $p, q > 1$ satisfying $\frac{1}{p} + \frac{1}{q} = 1$ are sufficient to obtain (III.4) and, for $p = q = 2$, equivalent to the generation of a polynomially bounded C_0-semigroup on Hilbert spaces.

If it is already known that A generates a C_0-semigroup, then it becomes much easier to check whether the semigroup is polynomially bounded at least for a large class of semigroups.

Following Eisner, Zwart [73], we say that an operator A has a *p-integrable resolvent* if for some/all $a, b > s_0(A)$ the conditions

$$\int_{-\infty}^{\infty} \|R(a + is, A)x\|^p \, ds < \infty \qquad \forall x \in X, \tag{III.6}$$

$$\int_{-\infty}^{\infty} \|R(b + is, A')y\|^q \, ds < \infty \qquad \forall y \in X' \tag{III.7}$$

hold, where $1 < p, q < \infty$ with $\frac{1}{p} + \frac{1}{q} = 1$. Note that in particular condition (I.11) from Subsection I.3.2 is satisfied for such semigroups by the Cauchy–Schwarz inequality.

Plancherel's theorem applied to the functions $t \mapsto e^{-at}T(t)x$ and $t \mapsto e^{-at}T^*(t)y$ for sufficiently large $a > 0$ implies that every generator of a C_0-semigroup on a Hilbert space has 2-integrable resolvent. Moreover, for generators on a Banach space with Fourier type $p > 1$ condition (III.6) is satisfied automatically. Finally, every generator of an analytic semigroup (in particular, every bounded operator) on an arbitrary Banach space has p-integrable resolvent for every $p > 1$. Intuitively, having p-integrable resolvent for some p is a good property of the generator A or/and a good property of the space X.

Theorem 1.20 (Eisner, Zwart [73]). *Let A be the generator of a C_0-semigroup $(T(t))_{t \geq 0}$ having p-integrable resolvent for some $p > 1$. Assume that $\mathbb{C}_0^+ = \{\lambda : \operatorname{Re} \lambda > 0\}$ is contained in the resolvent set of A and there exist $a_0 > 0$ and $M > 0$ such that the following conditions hold.*

(a) $\|R(\lambda, A)\| \leq \dfrac{M}{(\operatorname{Re} \lambda)^d}$ *for some $d \in [0, \infty)$ and all λ with $0 < \operatorname{Re} \lambda < a_0$;*

(b) $\|R(\lambda, A)\| \leq M$ *for all λ with $\operatorname{Re} \lambda \geq a_0$.*

Then $\|T(t)\| \leq K(1 + t^{2d-1})$ holds for some constant $K > 0$ and all $t \geq 0$.
Conversely, if $(T(t))_{t \geq 0}$ is a C_0-semigroup on a Banach space with

$$\|T(t)\| \leq K(1 + t^\gamma)$$

for some constants $\gamma \geq 0$, $K > 0$ and all $t \geq 0$, then for every $a_0 > 0$ there exists a constant $M > 0$ such that the resolvent of the generator satisfies conditions (a) and (b) above for $d = \gamma + 1$.

Proof. The second part of the theorem follows easily from the representation

$$R(\lambda, A)x = \int_0^\infty e^{-\lambda t} T(t) x \, dt.$$

The idea of the proof of the first part is based on the inverse Laplace transform representation of the semigroup presented in Subsection I.3.2 and the technique from Zwart [264] and Eisner, Zwart [72].

We first note that by conditions (a) and (b) we obtain $s_0(A) \leq 0$.

Next, since the function $\omega \mapsto R(a + i\omega, A)x$ is an element of $L^p(\mathbb{R}, X)$ for all $x \in X$, we conclude by the uniform boundedness theorem that there exists a constant $M_0 > 0$ such that

$$\|R(a + i\cdot, A)x\|_{L^p(\mathbb{R}, X)} \leq M_0 \|x\| \tag{III.8}$$

for all $x \in X$. Similarly, one obtains the dual result, i.e.,

$$\|R(b + i\cdot, A')y\|_{L^q(\mathbb{R}, X')} \leq \tilde{M}_0 \|y\| \tag{III.9}$$

for all $y \in X'$.

Take now $0 < r < a_0$. By the resolvent equality we have

$$R(r + i\omega), A)x = [I + (a - r)R(r + i\omega, A)] R(a + i\omega, A)x.$$

Hence

$$\|R(r + i\omega, A)x\| \leq [1 + |a - r| \|(R(r + i\omega, A)\|] \|R(a + i\omega, A)x\|$$

$$\leq \left[1 + |a - r| \frac{M}{r^d}\right] \|R(a + i\omega, A)x\|,$$

where we used (a). Combining this with the estimate (III.8), we find that

$$\|R(r+i\cdot,A)x\|_{L^p(\mathbb{R},X)} \le \left[1 + |a-r|\frac{M}{r^d}\right]M_0\|x\|$$

$$\le M_1\left[1 + \frac{1}{r^d}\right]\|x\|. \tag{III.10}$$

Similarly, we find that

$$\|R(r+i\cdot,A')y\|_{L^q(\mathbb{R},X')} \le \tilde{M}_1\left[1 + \frac{1}{r^d}\right]\|y\|. \tag{III.11}$$

From the estimates (III.10) and (III.11) we obtain

$$\int_{-\infty}^{\infty}|\langle R(r+i\omega,A)^2 x, y\rangle|\,d\omega$$

$$= \int_{-\infty}^{\infty}|\langle R(r+i\omega,A)x, R(r+i\omega,A)'y\rangle|\,d\omega$$

$$\le \|R(r+i\cdot,A)x\|_{L^p(\mathbb{R},X)}\|R(r+i\cdot,A')y\|_{L^q(\mathbb{R},X')}$$

$$\le M_1\tilde{M}_1\|x\|\|y\|\left[1+\frac{1}{r^d}\right]^2. \tag{III.12}$$

Convergence of the integral on the right-hand side of (III.12) implies that the inversion formula for the semigroup

$$T(t)x = \frac{1}{2\pi t}\int_{-\infty}^{\infty} e^{(r+is)t}R(r+is,A)^2 x\,ds$$

holds for all $x \in X$ by Theorem I.3.9. Notice that the condition $r > s_0(A)$ is essential. Combining this formula with (III.12) we obtain

$$|\langle T(t)x,y\rangle| \le \frac{1}{2\pi t}\int_{-\infty}^{\infty} e^{rt}|\langle R(r+i\omega,A)^2 x, y\rangle|\,d\omega$$

$$\le \frac{1}{2\pi t}e^{rt}M_1\tilde{M}_1\|x\|\|y\|\left[1+\frac{1}{r^d}\right]^2. \tag{III.13}$$

Since this holds for all $0 < r < a_0$, we may choose $r := \frac{1}{t}$ for t large enough giving

$$|\langle T(t)x,y\rangle| \le \frac{1}{2\pi t}eM_1\tilde{M}_1\|x\|\|y\|\left[1+t^d\right]^2. \tag{III.14}$$

So for large t the norm of the semigroup is bounded by Ct^{2d-1} for some constant C. Since any C_0-semigroup is uniformly bounded on compact time intervals, the result follows. $\square$

As mentioned above, every generator on a Hilbert space has 2-integrable resolvent, hence we have the following immediate corollary.

Corollary 1.21. *Let A generate a C_0-semigroup $(T(t))_{t\geq 0}$ on the Hilbert space H. If A satisfies conditions* (a) *and* (b) *of Theorem 1.20 for some $d \geq 0$ and $a_0 > 0$, then there exists $K > 0$ such that $\|T(t)\| \leq K[1 + t^{2d-1}]$ for all $t \geq 0$.*

Remark 1.22. Notice that conditions (a) and (b) for $0 \leq d < 1$ already imply $s_0(A) < 0$ (use the power series expansion for the resolvent). On the other hand, for generators with p-integrable resolvent the equality $w_0(T) = s_0(A)$ holds by Corollary 2.19. Combining these facts we obtain that in this case the semigroup is even uniformly exponentially stable. On the other hand, the exponential stability follows from the Theorem 1.20 only for $d < \frac{1}{2}$. So for $\frac{1}{2} \leq d < 1$ Theorem 1.20 does not give the best information about the growth of the semigroup. Nevertheless, for $d = 1$ the exponent $2d - 1$ given in Theorem 1.20 is best possible (see Eisner, Zwart [72]). For $d > 1$ this is not clear.

Note that the parameter $d = \gamma + 1$ in the converse implication of Theorem 1.20 is optimal for $\gamma \in \mathbb{N}$. Indeed, for $X := \mathbb{C}^n$ and

$$A := \begin{pmatrix} 0 & 1 & 0 & \dots & 0 \\ 0 & 0 & 1 & \dots & 0 \\ & & \vdots & & \\ 0 & 0 & 0 & \dots & 0 \end{pmatrix}$$

conditions (a) and (b) in Theorem 1.20 are fulfilled for $d = n$ and the semigroup generated by A grows exactly as t^{n-1}.

By Corollary 1.21 we see that the class of generators of polynomially bounded semigroups on a Hilbert space coincides with the class of generators of C_0-semigroups satisfying the resolvent conditions (a) and (b). For semigroups on Banach spaces this is not true since there exist C_0-semigroups with $w_0(T) > s_0(A)$ (see Engel, Nagel [78, Examples IV.3.2 and IV.3.3]).

As a corollary of Theorem 1.20 we have the following characterisation of polynomially bounded C_0-groups in terms of the resolvent of the generator.

Theorem 1.23. *Let A be the generator of a C_0-group $(T(t))_{t\in\mathbb{R}}$ on a Banach space. Assume that A has p-integrable resolvent for some $p > 1$. Then the group $(T(t))_{t\in\mathbb{R}}$ is polynomially bounded if and only if the following conditions on the operator A are satisfied.*

(a) $\sigma(A) \subset i\mathbb{R}$.

(b) *There exist $a_0 > 0$ and $d \geq 0$ such that $\|R(\lambda, A)\| \leq \dfrac{M}{|\operatorname{Re}\lambda|^d}$ for some constant M and all λ with $0 < |\operatorname{Re}\lambda| < a_0$.*

(c) $R(\lambda, A)$ *is uniformly bounded on* $\{\lambda : |\operatorname{Re}\lambda| \geq a_0\}$.

Proof. It is enough to show that the operator $-A$ also has p-integrable resolvent whenever A satisfies (a)–(c). Take any $a > 0$. Then by (b) or (c), respectively, $R(\lambda, A)$ is bounded on the vertical line $-a + i\mathbb{R}$. By the resolvent equation we obtain

$$\|R(-a + is, A)x\| \leq [1 + 2a\|R(-a + is, A)\|]\|R(a + is, A)x\|,$$

and therefore the function $s \mapsto \|R(-a + is, A)x\|$ also belongs to $L^p(\mathbb{R})$. The rest follows immediately from Theorem 1.20. $\qquad\square$

Again, this yields a characterisation of polynomially bounded C_0-groups on Hilbert spaces. Note that the relation between the growth of the group and the growth of the resolvent appearing in (b) of Theorem 1.23 is as in Theorem 1.20.

2 Uniform exponential stability

In this section we study the concept of uniform exponential stability of C_0-semigroups. Due to the unboundedness of the generator, this turns out to be more difficult than the corresponding notion for operators (compare Proposition II.1.3).

Uniform exponential stability is defined as follows.

Definition 2.1. A C_0-semigroup $T(\cdot)$ is called *uniformly exponentially stable* if there exist $M \geq 1$ and $\varepsilon > 0$ such that

$$\|T(t)\| \leq Me^{-\varepsilon t} \quad \text{for all } t \geq 0,$$

or, equivalently, if $\omega_0(T) < 0$.

For C_0-semigroups on finite-dimensional Banach spaces the classical Lyapunov theorem gives a simple characterisation in terms of the spectrum of the generator: A matrix semigroup $(e^{tA})_{t\geq 0}$ is uniformly exponentially stable if and only if $\text{Re}\,\lambda < 0$ for every $\lambda \in \sigma(A)$, i.e., if $s(A) < 0$. However, on infinite-dimensional spaces this equivalence is no longer valid.

The aim of this section is to characterise uniformly exponentially stable C_0-semigroups on Banach and Hilbert spaces, preferably through spectral and resolvent properties of the generator.

2.1 Spectral characterisation

The following elementary description of uniformly exponentially stable C_0-semigroups is the basis for many further results in this section, see Engel, Nagel [78, Prop. V.1.7].

Theorem 2.2. *For a C_0-semigroup $T(\cdot)$ on a Banach space X the following assertions are equivalent.*

(i) $T(\cdot)$ *is uniformly exponentially stable, i.e., $\omega_0(T) < 0$.*

(ii) $\lim\limits_{t\to\infty} \|T(t)\| = 0$.

(iii) $\|T(t_0)\| < 1$ *for some* $t_0 > 0$.

(iv) $r(T(t_0)) < 1$ *for some* $t_0 > 0$.

(v) $r(T(t)) < 1$ *for all* $t > 0$.

The proof of the non-trivial implication (iv)$\Rightarrow$(i) is based on the formula $r(T(t)) = e^{t\omega_0(T)}$.

This theorem shows that, in particular, stability in the norm topology already implies uniform exponential stability. Furthermore, the equivalences (i)$\Leftrightarrow$(iv)$\Leftrightarrow$(v) show that in order to check uniform exponential stability it suffices to determine the spectral radius of $T(t_0)$ for some t_0. Since, in most cases, the semigroup is unknown, this is not very helpful in general. However, if the spectrum of the semigroup, and therefore the spectral radius $r(T(t))$ can be obtained via some spectral mapping theorem from the spectrum of the generator, the analogue of the Lyapunov theorem holds. We devote the next section to this topic.

2.2   Spectral mapping theorems

The question is how to determine the spectrum and the growth bound of the semigroup using information about the generator only.

For bounded generators we know that the semigroup is given as the exponential function e^{tA} (via the Taylor series or the Dunford functional calculus), and the spectral mapping theorem

$$\sigma(T(t)) = e^{t\sigma(A)} \quad \text{for every } t \geq 0 \qquad \text{(III.15)}$$

holds. This combined with $r(T(t)) = e^{t\omega_0(T)}$ implies the equality

$$s(A) = \omega_0(T) \qquad \text{(III.16)}$$

and allows one to check the uniform exponential stability by determining the spectrum of the generator. It is well-known, see from e.g. Engel, Nagel [78, Examples IV.2.7, IV.3.3-4], neither (III.15) nor (III.16) holds in general.

We give a short survey of why and how the spectral mapping theorem (III.15) fails, and under which assumptions and modifications it becomes true. As a first modification, we should exclude the point 0 in (III.15) since it can occur in the left-hand, but not in the right-hand term. So we choose the following terminology.

Definition 2.3. A C_0-semigroup $T(\cdot)$ on a Banach space X with generator A satisfies *the spectral mapping theorem* (or has *the spectral mapping property*) if

$$\sigma(T(t)) \setminus \{0\} = e^{t\sigma(A)} \quad \text{for every } t \geq 0. \qquad \text{(III.17)}$$

For semigroups satisfying the spectral mapping theorem, equality (III.16) holds automatically by the formula $r(T(t)) = e^{t\omega_0(T)}$.

Note that the spectral mapping theorem always holds for the point and residual spectrum, see Engel, Nagel [78, Theorem IV.3.7]. Moreover, the spectral mapping theorem holds for certain quite large and important classes of C_0-semigroups.

Theorem 2.4. (see Engel, Nagel [78, Cor. IV.3.12]) *The spectral mapping theorem holds for all eventually norm continuous C_0-semigroups, hence for the following classes of semigroups:*

(i) *uniformly continuous semigroups (or, equivalently, for semigroups with bounded generators),*

(ii) *eventually compact semigroups,*

(iii) *analytic semigroups,*

(iv) *eventually differentiable semigroups.*

In particular, for any such semigroup equality (III.16) holds, implying that the semigroup is uniformly exponentially stable if and only if $s(A) < 0$.

For the proof of the first assertion we refer to Engel, Nagel [78, Cor. IV.3.12]. Equality (III.16) is then a consequence of the formula $r(T(t)) = e^{t\omega_0(T)}$.

However, the following example shows that even for simple multiplication semigroups the spectral mapping theorem can fail.

Example 2.5. Let $T(\cdot)$ be the multiplication semigroup on l^2 given by

$$T(x_1, x_2, \ldots) = (e^{tq_1} x_1, e^{tq_2} x_2, \ldots)$$

for $q_n := \frac{1}{n} + in$. Then

$$1 \in \sigma(T(2\pi)) = \overline{\{e^{2\pi/n} : n \in \mathbb{N}\}},$$

while

$$1 \notin e^{2\pi\sigma(A)} = \{e^{2\pi/n} : n \in \mathbb{N}\}.$$

So the spectral mapping theorem (III.17) does not hold.

We now look for some weaker forms of a spectral mapping theorem still implying (III.16). The first is motivated by the above example.

Definition 2.6. A C_0-semigroup $T(\cdot)$ with generator A satisfies *the weak spectral mapping theorem* (or has *the weak spectral mapping property*) if

$$\sigma(T(t)) \setminus \{0\} = \overline{e^{t\sigma(A)}} \setminus \{0\} \quad \text{for every } t \geq 0. \tag{III.18}$$

Every multiplication semigroup on spaces $C_0(\Omega)$ and $L^p(\Omega, \mu)$ satisfies the weak spectral mapping theorem, see Engel, Nagel [78, Prop. IV.3.13]. Consequently, we obtain (by the spectral theorem) that every semigroup of normal operators on a Hilbert space has the weak spectral mapping property. Moreover, the weak spectral mapping theorem holds for bounded C_0-groups on Banach spaces, see Engel, Nagel [78, Prop. IV.3.13], and, more generally, for every C_0-group with so-called non-quasianalytic growth, see Huang [131] and Huang, Nagel [197] for details.

We finally present another weaker spectral mapping property, still sufficient for equation (III.16).

Definition 2.7. A C_0-semigroup $T(\cdot)$ with generator A satisfies *the weak circular spectral mapping theorem* (or has *the weak circular spectral mapping property*) if

$$\Gamma \cdot \sigma(T(t)) \setminus \{0\} = \Gamma \cdot \overline{e^{t\sigma(A)}} \setminus \{0\} = \overline{e^{t\sigma(A)+i\mathbb{R}}} \setminus \{0\} \quad \text{for every } t \geq 0. \quad \text{(III.19)}$$

We thus obtain the following.

Proposition 2.8. *For a C_0-semigroup $T(\cdot)$ with generator A satisfying the weak circular spectral mapping theorem, the equality*

$$\omega_0(T) = s(A)$$

holds. In particular, $T(\cdot)$ is uniformly exponentially stable if and only if $s(A) < 0$.

The proof is again based on the formula $r(T(t)) = e^{t\omega_0(T)}$.

For concrete examples of semigroups satisfying the weak circular spectral mapping theorem see e.g. Greiner, Schwarz [111], Kramar, Sikolya [153] and Bátkai, Eisner, Latushkin [19].

Remark 2.9. In order to check the weak circular or some other spectral mapping theorem, a result due to Greiner is very useful, see Nagel (ed.) [196, Theorems A-III.7.8 and 7.10]. It states that, for a C_0-semigroup $T(\cdot)$ and its generator A, $e^{t\lambda} \in \rho(T(t))$ if and only if the resolvent of A exists and is Cesàro bounded on $\Lambda := \{\lambda + \frac{2\pi k}{t} : k \in \mathbb{Z}\}$. Moreover, on Hilbert spaces one can replace Cesàro boundedness by boundedness of the resolvent, see Gearhart [94] and Prüss [215]. For the evolution semigroup version of the spectral mapping theorem see Chicone, Latushkin [47, Section 2.2.2] or van Neerven [204, Section 2.5].

Note finally that spectral mapping theorems give much more information than just on uniform exponential stability. For example, they allow us to characterise hyperbolicity, periodicity, so-called Lyapunov exponents and so on, see e.g. Chicone, Latushkin [47, Sections 2.1.4, 6.2.3, 6.3.3] and Engel, Nagel [78, Sections IV.2.c, V.1.c].

2.3 Theorem of Datko–Pazy

The main result of this subsection is a classical theorem characterising uniform exponential stability of a C_0-semigroup in terms of the integrability of its orbits.

Theorem 2.10 (Datko–Pazy). *A C_0-semigroup $T(\cdot)$ on a Banach space X is uniformly exponentially stable if and only if, for some $p \in [1, \infty)$,*

$$\int_0^\infty \|T(t)x\|^p \, dt < \infty \quad \text{for all } x \in X. \tag{III.20}$$

Proof. Clearly, uniform exponential stability implies condition (III.20). We now prove the converse direction.

Assume that condition (III.20) holds. We first show that the semigroup $T(\cdot)$ is bounded. Take $M \geq 1$ and $\omega > 0$ satisfying $\|T(t)\| \leq Me^{\omega t}$ for every $t \geq 0$. Then

$$\frac{1 - e^{-p\omega t}}{p\omega}\|T(t)x\|^p = \int_0^t e^{-p\omega s}\|T(s)T(t-s)x\|^p \, ds$$

$$\leq M^p \int_0^t \|T(t-s)x\|^p \, ds \leq M^p \int_0^\infty \|T(s)x\|^p \, ds$$

holds for every $x \in X$ and $t \geq 0$, where the right-hand side is finite by assumption. Hence, every orbit $\{T(t)x : t \geq 0\}$ is bounded and therefore the semigroup $T(\cdot)$ is bounded by the uniform boundedness principle, hence $L := \sup_{t\geq 0}\|T(t)\| < \infty$.

We further observe that

$$t\|T(t)x\|^p = \int_0^t \|T(t-s)T(s)x\|^p \, ds \leq L^p \int_0^\infty \|T(s)x\|^p \, ds$$

for every $x \in X$ and $t \geq 0$. This implies boundedness of the set $\{t^{\frac{1}{p}}T(t)x, t > 0\}$ for every $x \in X$. By the uniform boundedness principle, $\sup_{t\geq 0} t^{\frac{1}{p}}\|T(t)\| < \infty$ and therefore $T(\cdot)$ is uniformly exponentially stable by Theorem 2.2 (iii). $\quad\square$

Datko [55] proved this theorem for $p = 2$ and Pazy [209, Theorem 4.4.1] extended it to the case $p \geq 1$.

There are various generalisations of the Datko–Pazy theorem. As an example we state theorems due to van Neerven and Rolewicz, see van Neerven [204, Corollary 3.1.6 and Theorem 3.2.2].

Theorem 2.11. *Let $T(\cdot)$ be a C_0-semigroup on a Banach space X, $p \in [1, \infty)$ and $\beta \in L^1_{loc}(\mathbb{R}_+)$ a positive function satisfying*

$$\int_0^\infty \beta(t)dt = \infty.$$

If

$$\int_0^\infty \beta(t)\|T(t)x\|^p \, dt < \infty \quad \text{for all } x \in X,$$

then $T(\cdot)$ is uniformly exponentially stable.

Theorem 2.12 (Rolewicz). *Let $T(\cdot)$ be a C_0-semigroup on a Banach space X. If there exists a strictly positive increasing function ϕ on $\mathbb{R}_+$ such that*

$$\int_0^\infty \phi(\|T(t)x\|)\, dt < \infty \quad \text{for all } x \in X,\ \|x\| \leq 1,$$

then $T(\cdot)$ is uniformly exponentially stable.

For further discussion of the above result we refer to van Neerven [204, pp. 110–111].

The following theorem is a weak version of the Datko–Pazy theorem due to Weiss [255].

Theorem 2.13 (Weiss). *Let $T(\cdot)$ be a C_0-semigroup on a Hilbert space X. If for some $p \in [1, \infty)$,*

$$\int_0^\infty |\langle T(t)x, y\rangle|^p\, dt < \infty \quad \text{for all } x \in X \text{ and } y \in X',$$

then $T(\cdot)$ is uniformly exponentially stable.

We refer to Weiss [256] for a discrete version of the above result.

Remark 2.14. The assertion of Theorem 2.13 also holds for bounded semigroups on general Banach spaces, see van Neerven [204, Theorem 4.6.3]. However, without the boundedness assumption it fails and $\omega_0(T) > 0$ becomes possible even for positive semigroups, see van Neerven [204, Theorem 4.6.5 and Example 1.4.4]. We refer to Tomilov [242] and van Neerven [204, Section 4.6] for other generalisations of Theorem 2.13.

For further generalisations of the Datko–Pazy theorem see, e.g., Vũ [252], van Neerven [204, Sections 3.3-4] and Theorem 3.9 below.

2.4   Theorem of Gearhart

For C_0-semigroups on Hilbert spaces there is a useful characterisation of uniform exponential stability in terms of the generator's resolvent on the right half plane. This is the classical theorem of Gearhart [94] and can be considered as a generalisation of the finite-dimensional Lyapunov theorem to the infinite-dimensional case. To this theorem and its generalisations this subsection is dedicated.

We begin with the following result on uniform exponential stability using the inverse Laplace transform method presented in Subsection I.3.2.

Theorem 2.15. *Let A generate a C_0-semigroup $T(\cdot)$ on a Banach space X such that the resolvent of A is bounded on the right half plane $\{z : \operatorname{Re} z > 0\}$. Assume further that*

$$\int_{-\infty}^\infty |\langle R^2(is, A)x, y\rangle|\, ds < \infty \quad \text{for all } x \in X \text{ and } y \in X'. \qquad \text{(III.21)}$$

Then $T(\cdot)$ *is uniformly exponentially stable.*

Proof. We first observe that, since the resolvent is bounded on the right half plane, it is also bounded on a half plane $\{z : \operatorname{Re} z > -\delta\}$ for some $\delta > 0$ by the power series expansion of the resolvent and hence $s_0(A) < 0$. Therefore, by condition (III.21) and Theorem I.3.8 we have

$$\langle T(t)x, y \rangle = \frac{1}{2\pi t} \int_{-\infty}^{\infty} e^{ist} \langle R^2(is, A)x, y \rangle \, ds \qquad \text{for all } x \in X, \ y \in X'.$$

The uniform boundedness principle now implies

$$|\langle T(t)x, y \rangle| \leq \frac{1}{2\pi t} \int_{-\infty}^{\infty} |\langle R^2(is, A)x, y \rangle| \, ds \leq \frac{1}{2\pi t} M \|x\| \|y\|$$

for some constant M and all $x \in X$, $y \in X'$, hence $\|T(t)\| \leq \frac{M}{2\pi t} \longrightarrow 0$ as $t \to \infty$. □

Note that this result generalises a result of Xu and Feng [259].

We now present Gearhart's theorem for which we give two proofs: one based on the theorem above, i.e., on the properties of the inverse Laplace transform, and the second using the Datko-Pazy theorem.

Theorem 2.16 (Gearhart, 1978). *Let A generate a C_0-semigroup $T(\cdot)$ on a Hilbert space H. Then $T(\cdot)$ is uniformly exponentially stable if and only if there exists a constant $M > 0$ such that*

$$\|R(\lambda, A)\| < M \quad \text{for all } \lambda \text{ with } \operatorname{Re} \lambda > 0. \tag{III.22}$$

Proof I. As in the proof of Theorem 2.15 we see, by the power series expansion for the resolvent, that $s_0(A) < 0$ holds. By Theorem 2.15 it suffices to check condition (III.21).

Take $a > \omega_0(T)$ and $x, y \in H$. By the Laplace transform representation of the resolvent of A and Plancherel's theorem applied to the function $t \mapsto e^{-at}T(t)x$ we obtain

$$\int_{-\infty}^{\infty} \|R(a + is, A)x\|^2 \, ds = \int_{0}^{\infty} e^{-2at} \|T(t)x\|^2 \, dt < \infty.$$

Further, by (III.22) and the resolvent identity, the estimate

$$\|R(is, A)x\| = \|[I + aR(is, A)]R(a + is, A)x\| \leq [1 + aM] \|R(a + is, A)x\|$$

holds for every $s \in \mathbb{R}$ and hence

$$\int_{-\infty}^{\infty} \|R(is, A)x\|^2 \, ds < \infty.$$

Applying the same arguments to the operator A^* and the function $t \mapsto T^*(t)y$ we obtain

$$\int_{-\infty}^{\infty} |\langle R^2(is, A)x, y\rangle| \, ds = \int_{-\infty}^{\infty} |\langle R(is, A)x, R(-is, A^*)y\rangle| \, ds$$

$$\leq \left(\int_{-\infty}^{\infty} \|R(is, A)x\|^2 \, ds\right)^{\frac{1}{2}} \left(\int_{-\infty}^{\infty} \|R(is, A^*)y\|^2 \, ds\right)^{\frac{1}{2}} < \infty,$$

and the theorem is proved.   □

Proof II. (Weiss [255]) Assume $\omega_0 := \omega_0(T) \geq 0$ and take $x \in H$ and a, a_0 with $0 \leq \omega_0 < a < a_0$. By the resolvent equation and (III.22) we have

$$\|R(a + is, A)x\| = \|(I + (a_0 - a)R(a + is, A))R(a_0 + is, A)x\|$$
$$\leq (1 + a_0 M)\|R(a_0 + is, A)x\| \quad \text{for all } s \in \mathbb{R}.$$

Therefore, again by the representation $R(a + is, A)x = \int_0^\infty e^{-(a+is)t}T(t)x \, dt$ and Plancherel's theorem, we have

$$\int_0^\infty e^{-2at}\|T(t)x\|^2 \, dt = \int_{-\infty}^\infty \|R(a + is, A)x\|^2 \, ds$$

$$\leq (1 + a_0 M)^2 \int_{-\infty}^\infty \|R(a_0 + is, A)x\|^2 \, ds.$$

Since this holds for every $a > \omega_0$, we conclude now by the monotone convergence theorem that

$$\int_0^\infty e^{-2\omega_0 t}\|T(t)x\|^2 \, dt < \infty \quad \text{for every } x \in X.$$

By the Datko–Pazy theorem applied to the rescaled semigroup $(e^{-\omega_0 t}T(t))_{t\geq 0}$, this implies uniform exponential stability for this semigroup, contradicting the definition of the growth bound.   □

By the representation $R(\lambda, A)x = \int_0^\infty e^{-\lambda t}T(t)x \, dt$, $\text{Re}\,\lambda > \omega_0(T)$, condition (III.22) is necessary for uniform exponential stability.

Remark 2.17. There are various other proofs of Gearhart's theorem, see, e.g., Prüss [215], Greiner in Nagel (ed.) [196, pp. 94–95], Engel, Nagel [78, pp. 302–303] or the original paper of Gearhart [94].

Remark 2.18. The boundedness of the resolvent on the right half plane in Gearhart's theorem cannot be replaced by the existence of the resolvent on the right half plane only. For an example of a semigroup on a Hilbert space satisfying $s(A) < s_0(A)$ see e.g. Engel, Nagel [78, Counterexample IV.3.4].

By the rescaling procedure one obtains the following corollary, see also Kaashoek and Verduyn Lunel [139] for a similar assertion (but a different method of proof).

Corollary 2.19. *Let A generate a C_0-semigroup $(T(t))_{t \geq 0}$ on a Banach space X. If for some $\delta > 0$ the integrability condition*

$$\int_{-\infty}^{\infty} |\langle R^2(a + is, A)x, y\rangle| \, ds < \infty \qquad \text{for all } x \in X, \ y \in X'$$

holds for all $s_0(A) < a < s_0(A) + \delta$, then $s_0(A) = \omega_0(T)$.

In particular, we see that for C_0-semigroups on Hilbert spaces the equality

$$s_0(A) = \omega_0(T)$$

holds. This is not true on Banach spaces. For an example of a C_0-semigroup satisfying $s_0(A) < \omega_0(T)$ see van Neerven [204, Example 4.2.9]. So the integrability of the resolvent on vertical lines in the above result cannot be omitted. It is interesting to know whether this condition can be weakened or what kind of other additional assumptions on the resolvent imply $s_0(A) = \omega_0(T)$.

Remark 2.20. One can apply Gearhart's theorem to positive C_0-semigroups on Hilbert lattices (as in e.g. Greiner, Nagel [110]): A positive C_0-semigroup with generator A on a Hilbert lattice is uniformly exponentially stable if and only if $[0, \infty) \subset \rho(A)$. This follows immediately from Gearhart's theorem and the fact that

$$\|R(a + is, A)\| \leq \|R(a, A)\| \quad \text{for all } s \in \mathbb{R}$$

holds for every $a > s(A)$, see e.g. Engel, Nagel [78, p.355].

For further generalisations of Gearhart's theorem see e.g. Herbst [125], Huang [132], Weis, Wrobel [258]. A Fourier multiplier approach to uniform exponential stability and hyperbolicity and a generalisation of Gearhart's theorem is in Latushkin, Shvydkoy [162] and Latushkin, Räbiger [161]. In addition, we refer to Chicone, Latushkin [47, Section II.2.2] for analogues of Gearhart's theorem in Banach spaces in the context of the evolution semigroups.

3 Strong stability

In the following we consider concepts weaker than uniform exponential stability and start with strong stability. This notion is not so well-understood and only in 1988 Arendt, Batty and Lyubich, Vũ obtained a simple sufficient spectral condition which, however, is far from being necessary. In 2001 Tomilov found a resolvent condition on the generator which is sufficient on Banach spaces and equivalent on Hilbert spaces. The condition is not so simple as the one in Gearhart's theorem and uses the second power of the resolvent of the generator on vertical lines.

3.1 Preliminaries

We first introduce strongly stable C_0-semigroups and then deduce some fundamental properties.

Definition 3.1. A C_0-semigroup $T(\cdot)$ on a Banach space X is called *strongly stable* if $\lim_{t\to\infty} \|T(t)x\| = 0$ for every $x \in X$.

The first example below is, in a certain sense, typical on Hilbert spaces, see Theorem 3.11 below.

Example 3.2. (a) (Shift semigroup). Consider $H := L^2(\mathbb{R}_+, H_0)$ for a Hilbert space H_0 and $T(\cdot)$ defined by

$$T(t)f(s) := f(s+t), \quad f \in H, \ t, s \geq 0. \tag{III.23}$$

The semigroup $T(\cdot)$ is called the *left shift semigroup* on H and is strongly stable. Note that the spectrum of its generator is the whole left halfplane.

The same semigroup on the spaces $C_0(\mathbb{R}_+, X)$ and $L^p(\mathbb{R}_+, X)$, X a Banach space, is also strongly stable for $1 \leq p < \infty$, but not for $p = \infty$.

(b) (Multiplication semigroup, see Engel, Nagel [78, p. 323]). Consider $X := C_0(\Omega)$ for a locally compact space Ω and the operator A given by

$$Af(s) := q(s)f(s), \quad f \in X, \ s \in \Omega,$$

with the maximal domain $D(A) := \{f \in X : qf \in X\}$, where q is a continuous function on Ω. The operator A generates the C_0-semigroup given by

$$T(t)f(s) = e^{tq(s)}f(s), \quad f \in X, \ s \in \Omega,$$

if and only if $\operatorname{Re} q$ is bounded from above. The semigroup is bounded if $\operatorname{Re} q(s) \leq 0$ for every $s \in \Omega$. Moreover, the semigroup is strongly stable if and only if $\operatorname{Re} q(s) < 0$ for every $s \in \Omega$. Indeed, if $\operatorname{Re} q(s) < 0$, then

$$\|T(t)f\| \leq \sup_{s \in K} e^{t \operatorname{Re} q(s)} \|f\| \xrightarrow[t\to\infty]{} 0$$

for every function f with compact support K. By the density of these functions $T(\cdot)$ is strongly stable. Conversely, if $q(s_0) \in i\mathbb{R}$, then $\|T(t)f\| \geq |f(s_0)|$ for every $f \in X$ and hence $T(\cdot)$ is not strongly stable.

Note that $\sigma(A) = \overline{q(\Omega)}$ and therefore every closed set contained in the closed left halfplane can occur as the spectrum of the generator of a strongly stable C_0-semigroup.

For a concrete example of strongly stable and strongly convergent semigroups appearing in mathematical models of biological processes see, e.g., Bobrowski [36], Bobrowski, Kimmel [37, 38].

The following property of strongly stable semigroups follows directly from the uniform boundedness principle.

Remark 3.3. Every strongly stable C_0-semigroup $T(\cdot)$ is bounded, hence $\sigma(A) \subset \{z : \operatorname{Re} z \leq 0\}$ holds for the generator A. Note that conditions $P_\sigma(A) \cap i\mathbb{R} = \emptyset$ and $P_\sigma(A') \cap i\mathbb{R} = \emptyset$ are also necessary for strong stability by the spectral mapping theorem for the point and residual spectrum, see Engel, Nagel [78, Theorem IV.3.7].

We now state an elementary but very useful property implying strong stability.

Lemma 3.4. *Let $T(\cdot)$ be a bounded C_0-semigroup on a Banach space X and let $x \in X$.*

(a) *If there exists an unbounded sequence $\{t_n\}_{n=1}^\infty \subset \mathbb{R}_+$ such that $\lim_{n \to \infty} \|T(t_n)x\| = 0$, then $\lim_{t \to \infty} \|T(t)x\| = 0$.*

(b) *If $T(\cdot)$ is a contraction semigroup, then $\lim_{t \to \infty} \|T(t)x\|$ exists.*

Proof. The second part follows from the fact that for contraction semigroups the function $t \mapsto \|T(t)x\|$ is non-increasing. To verify (a) assume that $\|T(t_n)x\| \to 0$ and take $\varepsilon > 0$, $n \in \mathbb{N}$ such that $\|T(t_n)x\| < \varepsilon$ and $M := \sup_{t \geq 0} \|T(t)\|$. We obtain

$$\|T(t)x\| \leq \|T(t - t_n)\| \|T(t_n)x\| < M\varepsilon$$

for every $t \geq t_n$, and (a) is proved. □

Remark 3.5. As in the discrete case (see Remark II.2.5), assertion (b) in the above lemma is not true for general bounded C_0-semigroups. Indeed, consider the Hilbert space $L^2(\mathbb{R}_+)$ endowed with the equivalent norm

$$\|f\| := \left(\|f \cdot \chi_1\|_2^2 + \frac{1}{4} \|f \cdot \chi_2\|_2^2 \right)^{\frac{1}{2}},$$

where χ_1 and χ_2 denote the characteristic functions of the sets $\bigcup_{n=0,1,2,\ldots}[2n, 2n+1]$ and $\bigcup_{n=0,1,2,\ldots}[2n+1, 2n+2]$, respectively. On this space take the right shift semigroup $T(\cdot)$ satisfying $\sup_{t \geq 0} \|T(t)\| = 2$. Then $\|T(2n)\chi_1\| = 1$ and $\|T(2n+1)\chi_1\| = \frac{1}{2}$ holds for every $n \in \mathbb{N}$ and hence

$$\frac{1}{2} = \liminf_{t \to \infty} \|T(t)\chi_1\| \neq \limsup_{t \to \infty} \|T(t)\chi_1\| = 1.$$

An immediate corollary of Lemma 3.4 is the following.

Corollary 3.6. *Let $T(\cdot)$ be a bounded C_0-semigroup on a Banach space X and $x \in X$. Then the following assertions are equivalent.*

(a) $\lim_{t \to \infty} \|T(t)x\| = 0$;

(b) $\displaystyle \lim_{t \to \infty} \frac{1}{t} \int_0^t \|T(s)x\|^p \, ds = 0$ *for some/all $p \geq 1$.*

In particular, $T(\cdot)$ is strongly stable if and only if (b) holds for every $x \in X$.

The following theorem, similar to Corollary 3.6, gives an equivalent description of strong stability without assuming boundedness, see Zwart [263] and Guo, Zwart [113].

Theorem 3.7. *For a C_0-semigroup $T(\cdot)$ on a Banach space X and $x \in X$, the following assertions are equivalent.*

(a) $\lim_{t\to\infty} T(t)x = 0$.

(b) *For some/all $p, q > 1$ with $\frac{1}{p} + \frac{1}{q} = 1$*

$$\lim_{t\to\infty} \frac{1}{t} \int_0^t \|T(s)x\|^p \, ds = 0,$$

$$\sup_{t\geq 0} \frac{1}{t} \int_0^t \|T'(s)y\|^q \, ds < \infty \quad \text{for all } y \in X'.$$

In particular, $T(\cdot)$ is strongly stable if and only if (b) *holds for all $x \in X$.*

The proof is an easy modification of the one of Theorem 1.6.

Remark 3.8. Note that for bounded C_0-semigroups the second part of condition (b) holds automatically.

We finally state a continuous analogue of the result of Müller (Theorem II.2.8) on the asymptotic behaviour of semigroups which are not uniformly exponentially stable, see van Neerven [204, Lemma 3.1.7].

Theorem 3.9 (van Neerven). *Let $T(\cdot)$ be a C_0-semigroup on a Banach space X with $\omega_0(T) \geq 0$. Then for every function $\alpha : \mathbb{R}_+ \to [0, 1)$ converging monotonically to 0 there exists $x \in X$ with $\|x\| = 1$ such that*

$$\|T(t)x\| \geq \alpha(t) \quad \text{for every } t \geq 0.$$

Theorem 3.9 means that strongly stable semigroups which are not exponentially stable possess arbitrarily slowly decreasing orbits.

Strong stability of semigroups on Hilbert spaces can be characterised in terms of strong stability of its cogenerator, see Subsection V.2.5. A necessary and sufficient condition for strong stability on Hilbert spaces using the resolvent of the generator is given in Subsection 3.4 below.

3.2 Representation as shift semigroups

We now give a representation of strongly stable semigroups following directly from Theorem 1.9.

Proposition 3.10. (a) *Every strongly stable contraction semigroup on a Banach space X is isometrically isomorphic to the left shift semigroup on a closed subspace of $C_0(\mathbb{R}_+, X)$.*

(b) *Every strongly stable C_0-semigroup on a Banach space X is isomorphic to the left shift semigroup on a closed subspace of $C_0(\mathbb{R}_+, X_1)$, where $X_1 = X$ endowed with an equivalent norm.*

In analogy to Theorem II.2.11, Lax and Phillips showed that Example 3.2 (a) represents the general situation for contractive strongly stable C_0-semigroups on Hilbert spaces.

Theorem 3.11 (Lax, Phillips [164, p. 67], see also Lax [163, pp. 450–451]). *Let $T(\cdot)$ be a strongly stable contraction semigroup on a Hilbert space H. Then $T(\cdot)$ is unitarily isomorphic to a left shift, i.e., there is a Hilbert space H_0 and a unitary operator $U : H \to H_1$ for some closed subspace $H_1 \subset L^2(\mathbb{R}_+, H_0)$ such that $UT(\cdot)U^{-1}$ is the left shift on H_1.*

Proof. The idea is again to identify a vector x with its orbit $s \mapsto T(s)x$.

Strong stability of $T(\cdot)$ implies the equality

$$\|x\|^2 = -\int_0^\infty \frac{d}{ds}(\|T(s)x\|^2)\,ds = \int_0^\infty (-2\mathrm{Re}\,\langle AT(s)x, T(s)x \rangle)\,ds \qquad (\text{III}.24)$$

for every $x \in D(A)$.

Define now the new seminorm

$$\|x\|_Y^2 := -2\mathrm{Re}\,\langle Ax, x \rangle \quad \text{on } D(A),$$

which is non-negative by the Lumer–Phillips theorem. Note that it comes from the scalar semiproduct $\langle x, y \rangle_Y := -\langle Ax, y \rangle - \langle x, Ay \rangle$. Consider the subspace $H_0 := \{x \in D(A) : \|x\|_Y = 0\}$ with its completion $Y := (D(A)/H_0, \|\cdot\|_Y)\tilde{\;}$.

We now define the operator $J : H \to L^2(Y)$ by

$$(Jx)(s) := T(s)x, \quad x \in H, \ s \geq 0,$$

which is an isometry by equality (III.24). Therefore it is unitary from H to its (closed) range. Since the semigroup $(JT(t)J^{-1})_{t \geq 0}$ is the right shift semigroup on rg J, the proof is complete. $\qquad\square$

Note that in contrast to the analogous Proposition 3.10, the above representation respects the Hilbert space structure.

3.3 Spectral conditions

In this subsection we discuss spectral conditions on the generator implying strong stability of the semigroup. Analogously to the discrete case, these conditions use "smallness" of the spectrum of the generator on the imaginary axis.

The first result is the stability theorem for C_0-semigroups proved by Arendt, Batty [9] and Lyubich, Vũ [180] independently, generalising a result of Sklyar, Shirman [234] on semigroups with bounded generators. The discrete version of this result has been treated in Subsection II.2.3.

Theorem 3.12 (Arendt–Batty–Lyubich–Vũ, 1988). *Let $T(\cdot)$ be a bounded semigroup on a Banach space X with generator A. Assume that*

(i) $P_\sigma(A') \cap i\mathbb{R} = \emptyset$;

(ii) $\sigma(A) \cap i\mathbb{R}$ *is countable.*

Then $T(\cdot)$ is strongly stable.

The proof based on the Lyubich and Vũ construction of the *isometric limit semigroup*, which is a natural modification of the discrete case, can be found in Engel, Nagel [78, pp. 263–264 and Theorem V.2.21]. For the alternative proof by Arendt and Batty using the Laplace transform see Arendt, Batty [9] or Arendt, Batty, Hieber, Neubrander [10, Theorem 5.5.5].

For generalisations and extensions of Theorem 3.12 see Batty, Vũ [27], Batty [22], Batty, van Neerven, Räbiger [26]. For a concrete application of the above result to semigroups arising in biology see Bobrowski, Kimmel [37].

Remark 3.13. 1) Conditions (i) and (ii) in the above theorem are of different nature. The first one is necessary for strong stability while the second is a useful, but restrictive assumption.

2) For weakly relatively compact semigroups, in particular for bounded semigroups on reflexive Banach spaces, condition (i) is equivalent to

$$P_\sigma(A) \cap i\mathbb{R} = \emptyset$$

by the mean ergodic theorem, see Subsection I.1.6.

An immediate corollary of the above theorem is the following simple spectral criterion (see also Chill, Tomilov [50] for an alternative proof using a theorem of Ingham on the Laplace transform).

Corollary 3.14. *Let A generate a bounded C_0-semigroup $T(\cdot)$ on a Banach space. If $\sigma(A) \cap i\mathbb{R} = \emptyset$, then $T(\cdot)$ is strongly stable.*

Proof. The assertion follows from $P_\sigma(A') \cap i\mathbb{R} = R_\sigma(A) \cap i\mathbb{R} = \emptyset$ and the Arendt–Batty–Lyubich–Vũ theorem.   □

Theorem 3.12 is particularly useful if one combines it with the Perron–Frobenius spectral theory for positive semigroups on Banach lattices (see Nagel (ed.) [196] and Schaefer [227]).

Corollary 3.15. *Let A generate a bounded, eventually norm continuous, positive semigroup $T(\cdot)$ on a Banach lattice X. Then $T(\cdot)$ is strongly stable if and only if $0 \notin P_\sigma(A')$.*

Proof. By the Perron–Frobenius theory, the spectrum of the generator of a positive bounded semigroup on the imaginary axis is additively cyclic, i.e., $i\alpha \in \sigma(A)$ implies $i\alpha\mathbb{Z} \in \sigma(A)$ for real α, see Nagel (ed.) [196, Theorem C-III.2.10 and Proposition C-III.2.9]. Moreover, the spectrum of the generator of an eventually norm continuous semigroup is bounded on every vertical line, see Engel, Nagel [78, Theorem II.4.18]. The combination of these two facts leads to $\sigma(A) \cap i\mathbb{R} \subset \{0\}$. The theorem of Arendt–Batty–Lyubich–Vũ concludes the argument. $\qquad\square$

Remark 3.16. We note that in the above result the condition $0 \notin P_\sigma(A')$ cannot be replaced by $0 \notin P_\sigma(A)$. This can be easily seen for $A := T_r - I$ on l^1, where T_r is the right shift operator on l^1.

Moreover, the following is a further direct corollary of Theorem 3.12 and the mean ergodic theorem.

Corollary 3.17. *Let $T(\cdot)$ be a bounded holomorphic C_0-semigroup on a Banach space with generator A. Then $T(\cdot)$ is strongly stable if and only if $0 \notin P_\sigma(A')$.*

For an alternative proof see Bobrowski [35].

As in the discrete case, the theorem of Arendt–Batty–Lyubich–Vũ can be extended to completely non-unitary contraction semigroups on Hilbert spaces.

Theorem 3.18. (see Foiaş and Sz.-Nagy [238, II.6.7] and Kérchy, van Neerven [148]) *Let $T(\cdot)$ be a completely non-unitary contractive C_0-semigroup on a Hilbert space with generator A. If*

$$\sigma(A) \cap i\mathbb{R} \quad \text{has Lebesgue measure } 0,$$

then $T(\cdot)$ and $T^*(\cdot)$ are both strongly stable.

See also Kérchy, van Neerven [148] for related results.

The following example shows that in the above theorem one cannot replace a completely non-unitary by a contractive semigroup with $P_\sigma(A) \cap i\mathbb{R} = \emptyset$. The idea of this elegant construction belongs to Dávid Kunszénti-Kovács (oral communication).

Example 3.19. Take the Cantor set C on $[0,1]$ constructed from the intervals $(a_1, b_1) = (\frac{1}{3}, \frac{2}{3})$, $(a_2, b_2) = (\frac{1}{9}, \frac{2}{9})$, $(a_3, b_3) = (\frac{7}{9}, \frac{8}{9})$ and so on. Take further the Cantor set $\tilde{C}$ constructed from the intervals $\{(\tilde{a}_n, \tilde{b}_n)\}$ such that $\tilde{C}$ has measure $\frac{1}{2}$. By the natural linear transformation mapping each $(\tilde{a}_n, \tilde{b}_n)$ onto (a_n, b_n) and its continuation we obtain a bijective monotone map $j : \tilde{C} \to C$ being an analogue of the Cantor step function. The image of the Lebesgue measure μ under j we denote by μ_j. We observe that by construction μ_j is continuous. Note that, since $\mu_j(C) = \frac{1}{2}$, it is not absolutely continuous.

Consider $H := L^2(C, \mu_j)$ and the bounded operator A on H defined by $Af(s) := isf(s)$. Note that the corresponding unitary group is defined by $T(t)f(s) = e^{ist}f(s)$. Since μ_j is continuous, $P_\sigma(A') = P_\sigma(A) = \emptyset$. Moreover, $\sigma(A) = iC$ and therefore $\mu(\sigma(A)) = 0$. However, $T(\cdot)$ is unitary and hence $\lim_{t\to\infty} \|T(t)x\| = 0$ or $\lim_{t\to\infty} \|T^*(t)x\| = 0$ implies $x = 0$.

We finish this subsection by the observation that $\sigma(A) \cap i\mathbb{R}$ for the generator A of a strongly stable semigroup can be an arbitrary closed subset of $i\mathbb{R}$, see Example 3.2(b). So, "smallness" conditions on the boundary spectrum are far from being necessary for strong stability.

3.4　Characterisation via resolvent

We now present a powerful resolvent approach to strong stability introduced by Tomilov [243]. In contrast to the spectral approach from the previous subsection, resolvent conditions are necessary and sufficient at least for C_0-semigroups on Hilbert spaces.

Theorem 3.20 (Tomilov). *Let A generate a C_0-semigroup $T(\cdot)$ on a Banach space X satisfying $s_0(A) \leq 0$ and $x \in X$. Consider the following assertions.*

(a) $\displaystyle \lim_{a \to 0+} a \int_{-\infty}^{\infty} \|R(a + is, A)x\|^2 \, ds = 0,$

$\displaystyle \limsup_{a \to 0+} a \int_{-\infty}^{\infty} \|R(a + is, A')y\|^2 \, ds < \infty$ *for all $y \in X'$;*

(b) $\displaystyle \lim_{t \to \infty} \|T(t)x\| = 0.$

Then (a) *implies* (b). *Moreover, if X is a Hilbert space, then* (a)$\Leftrightarrow$(b).

In particular, condition (a) *for all $x \in X$ implies strong stability of $T(\cdot)$ and, on Hilbert spaces, is equivalent to it.*

Proof. We first prove that (a)$\Rightarrow$(b). By Theorem I.3.9 (see also Theorem I.3.8) and the Cauchy-Schwarz inequality we have

$$|\langle T(t)x, y \rangle| \leq \frac{e^{at}}{2\pi t} \int_{-\infty}^{\infty} |\langle R^2(a + is, A)x, y \rangle| \, ds$$

$$\leq \frac{e^{at}}{2\pi t} \left(\int_{-\infty}^{\infty} \|R(a + is, A)x\|^2 \, ds \right)^{\frac{1}{2}} \left(\int_{-\infty}^{\infty} \|R(a + is, A')y\|^2 \, ds \right)^{\frac{1}{2}}$$

for every $t > 0$, $a > 0$ and $y \in X'$. By (a) and the uniform boundedness principle there exists a constant $M > 0$ such that

$$a \int_{-\infty}^{\infty} \|R(a + is, A')y\|^2 \, ds \leq M^2 \|y\|^2 \quad \text{for every } y \in X' \text{ and } a > 0.$$

Therefore, we have

$$\|T(t)x\| \leq \frac{M e^{at}}{2\pi t a} \left(a \int_{-\infty}^{\infty} \|R(a + is, A)x\|^2 \, ds \right)^{\frac{1}{2}}. \tag{III.25}$$

By choosing $a := \frac{1}{t}$, we obtain by (III.25) $\lim_{t \to \infty} \|T(t)x\| = 0$.

Assume now that $T(\cdot)$ is strongly stable on a Hilbert space X. By Parseval's equality

$$a \int_{-\infty}^{\infty} \|R(a + is, A)x\|^2 \, ds = a \int_{0}^{\infty} e^{-2at} \|T(t)x\|^2 \, dt,$$

where the right-hand side is one half times the Abel mean of the function $t \mapsto \|T(t)x\|^2$. Therefore it converges to zero as $a \to 0+$ by the strong stability of $T(\cdot)$. This proves the first part of (a).

The second condition in (a) follows from Theorem 1.11.　　□

Remark 3.21. Corollary 2.19 shows that on Banach spaces the integrals appearing in (a) above do not converge in general.

We now show that in the above theorem (b) does not imply (a) in general even for bounded generators for which all the integrals in (a) converge. This is the following example which is a modification of Example 1.14.

Example 3.22. Take a bounded sequence $\{a_n\}_{n=1}^{\infty} \subset \{z : \operatorname{Re} z < 0\}$ such that $\overline{\{a_n : n \in \mathbb{N}\}} \cap i\mathbb{R} = i[-1, 1]$. Consider $X = l^1$ and the bounded operator A given by $A(x_1, x_2, \ldots) = (a_1 x_1, a_2 x_2, \ldots)$ on X. The semigroup generated by A is strongly stable by Example 3.2 (b).

The adjoint of A is given by $A'(x_1, x_2, \ldots) = (a_1 x_1, a_2 x_2, \ldots)$ on l^{∞}. So we have for every $a > 0$, $|b| < 1$ and $y \in U_{\frac{1}{2}}(1)$,

$$\|R(a + ib, A')y\| = \sup_{n \in \mathbb{N}} \frac{|x_n|}{|a + ib - a_n|} \geq \frac{1}{2 \operatorname{dist}(a + ib, i[-1, 1])} = \frac{1}{2a}.$$

Therefore we obtain

$$a \int_{-\infty}^{\infty} \|R(a + ib, A')y\|^2 \, db \geq a \int_{-1}^{1} \frac{1}{4a^2} \, db = \frac{1}{2a} \to \infty \quad \text{as } a \to 0+,$$

so the second part of (a) in Theorem 3.20 does not hold for y from the open set $U_{\frac{1}{2}}(1)$ in X'.

By Theorems 1.11 and 3.20 one immediately obtains the following characterisation of strong stability in the case of bounded C_0-semigroups on Hilbert spaces.

Corollary 3.23. *Let A generate a bounded semigroup $T(\cdot)$ on a Hilbert space H and $x \in X$. Then $\|T(t)x\| \to 0$ if and only if*

$$\lim_{a \to 0+} a \int_{-\infty}^{\infty} \|R(a + is, T)x\|^2 \, ds = 0. \tag{III.26}$$

In particular, $T(\cdot)$ is strongly stable if and only if (III.26) holds for every x in a dense set of H.

It is still an open question whether the above characterisation holds for C_0-semigroups on Banach spaces.

Remark 3.24. There are more results on strong stability of C_0-semigroups and their connection to the behaviour of the generator's resolvent. As some recent contributions we mention papers by Batkai, Engel, Prüss, Schnaubelt [20], Batty, Duyckaerts [25] and Borichev, Tomilov [39] on polynomial decay of the orbits and Chill, Tomilov [49] on characterisations of strong stability on Banach spaces with Fourier type. For so-called pointwise resolvent conditions see Tomilov [243] and Batty, Chill, Tomilov [24]. We recommend the excellent overview by Chill, Tomilov [50] for further results and references.

4 Weak stability

In this section we consider stability of C_0-semigroups with respect to the weak operator topology. As in the discrete case, this turns out to be much more difficult than its strong and uniform analogues.

4.1 Preliminaries

We begin with the definition and some examples.

Definition 4.1. A C_0-semigroup $T(\cdot)$ on a Banach space X is called *weakly stable* if $\lim_{t\to\infty} \langle T(t)x, y \rangle = 0$ for every $x \in X$ and $y \in X'$.

Note that by the uniform boundedness principle every weakly stable semigroup $T(\cdot)$ on a Banach space is bounded, hence $\omega_0(T) \leq 0$ holds. In particular, the spectrum of the generator A belongs to the closed left half plane. Moreover, the spectral conditions $P_\sigma(A) \cap i\mathbb{R} = \emptyset$ and $R_\sigma(A) \cap i\mathbb{R} = P_\sigma(A') \cap i\mathbb{R} = \emptyset$ are necessary for weak stability.

Example 4.2. (a) The left and right shift semigroups are weakly stable and isometric (and hence not strongly stable) on the spaces $C_0(\mathbb{R}, X)$ and $L^p(\mathbb{R}, X)$ for every Banach space X and $1 < p < \infty$.

(b) The right shift semigroup on $L^p(\mathbb{R}_+, X)$ for a Banach space X defined by

$$(T(t)f)(s) = \begin{cases} f(s - t), & s \geq t, \\ 0, & s < t \end{cases}$$

is an isometric semigroup (hence not strongly stable) but is weakly stable for $1 < p < \infty$. When $p = 2$ and X is a Hilbert space, this semigroup is called the *(continuous) unilateral shift*, see, e.g., Sz.-Nagy and Foiaş [238, p. 150]. Note that the adjoint semigroup of the unilateral shift is the left shift on the same space and hence is strongly stable. Therefore, there is no subspace on which the restriction of a unilateral shift becomes unitary. We will see in Subsection 4.2 that unilateral shifts represent the general situation of isometric completely non-unitary weakly stable semigroups on Hilbert spaces.

(c) Consider $H := L^2(\mathbb{R})$ and the multiplication semigroup $T(\cdot)$ given by

$$T(t)f(s) := e^{tq(s)} f(s)$$

for some measurable function q with $\sup_{s \in \mathbb{R}} \operatorname{Re} q(s) < \infty$. Then $T(\cdot)$ is strongly stable if and only if $\operatorname{Re} q(s) < 0$ a.e., see Example 3.2 (b). If $\sup \operatorname{Re} q(s) \leq 0$, then $T(\cdot)$ is weakly stable if and only if $\lim_{t\to\infty} \int_a^b e^{tq(s)}\, ds = 0$ for every $[a, b] \subset \mathbb{R}$. This is the case for, e.g., $q(s) = ias^\beta$ for any $\alpha, \beta \in \mathbb{R} \setminus \{0\}$.

We now present a simple condition implying weak stability of C_0-semigroups using the following concept, cf. Definition II.3.3.

Definition 4.3. A set $M \subset \mathbb{R}_+$ is *relatively dense* or *syndetic* in $\mathbb{R}_+$ if there exists a number $\ell > 0$ such that M intersects every sub-interval of $\mathbb{R}_+$ of length ℓ.

It turns out that weak convergence to zero for such a sequence already implies weak stability.

Theorem 4.4. *Let $T(\cdot)$ be a C_0-semigroup and suppose that weak-$\lim_{n\to\infty} T(t_n) = 0$ for some relatively dense sequence $\{t_n\}_{j=1}^{\infty}$. Then $T(\cdot)$ is weakly stable.*

Proof. Without loss of generality, by passing to a subsequence if necessary, we assume that $\{t_n\}_{n=1}^{\infty}$ is monotone increasing and set $\ell := \sup_{n\in\mathbb{N}}(t_{n+1} - t_n)$, which is finite by assumption. Since every C_0-semigroup is bounded on compact time intervals and $(T(t_n))_{n\in\mathbb{N}}$ is weakly converging, hence bounded, we obtain that the semigroup $(T(t))_{t\geq 0}$ is bounded.

Fix $x \in X$, $y \in X'$. For $t \in [t_n, t_{n+1}]$ we have

$$\langle T(t)x, y\rangle = \langle T(t - t_n)x, T'(t_n)y\rangle,$$

where $(T'(t))_{t\geq 0}$ is the adjoint semigroup. We note that by assumption $T'(t_n)y \to 0$ in the weak*-topology.

Further, the set $K_x := \{T(s)x : 0 \leq s \leq \ell\}$ is compact in X and $T(t - t_n)x \in K_x$ for every $n \in \mathbb{N}$. Since pointwise convergence is equivalent to the uniform convergence on compact sets (see, e.g., Engel, Nagel [78, Prop. A.3]), we see that $\lim_{t\to\infty}\langle T(t)x, y\rangle = 0$. $\square$

Remark 4.5. Taking $t_n = n$ in the above theorem, we see that a C_0-semigroup $T(\cdot)$ is weakly stable if and only if the operator $T(1)$ is weakly stable. This builds a bridge between weak stability of discrete and continuous semigroups, and we will return to this aspect later in Section V.1.

Remark 4.6. One cannot drop the relative density assumption in Theorem 4.4 or even replace it by the assumption of density 1, see Section 5 below.

We finally mention that by Theorem II.3.8, weak orbits of a C_0-semigroup $T(\cdot)$ which is not exponentially stable can decrease arbitrary slowly. More precisely, for every positive function $a : \mathbb{R}_+ \to \mathbb{R}_+$ decreasing to zero, there exist $x \in X$ and $y \in X'$ such that the corresponding weak orbit satisfies $|\langle T(t_j)x, y\rangle| \geq a(t_j)$ for some sequence $\{t_j\}_{j=1}^{\infty}$ converging to infinity and all $j \in \mathbb{N}$. However, it is not clear whether the result of Badea, Müller [12] holds in the continuous case, i.e., whether one may replace $\{t_j\}_{j=1}^{\infty}$ by $\mathbb{R}_+$ for weakly stable semigroups.

4.2 Contraction semigroups on Hilbert spaces

In this subsection we present some classical theorems on the decomposition of contractive C_0-semigroups on Hilbert spaces with respect to their qualitative behaviour.

We begin with the decomposition into unitary and completely non-unitary parts due to Foiaş and Sz.-Nagy [237].

Theorem 4.7 (Foiaş, Sz.-Nagy). *Let $T(\cdot)$ be a contraction semigroup on a Hilbert space H. Then H is the orthogonal sum of two $T(\cdot)$- and $T^*(\cdot)$-invariant subspaces H_1 and H_2 such that*

(a) *H_1 is the maximal subspace on which the restriction $T_1(\cdot)$ of $T(\cdot)$ is unitary;*

(b) *the restrictions of $T(\cdot)$ and $T^*(\cdot)$ to H_2 are weakly stable.*

We present the proof of Foguel given in [86] being analogous to the proof of the discrete version of this result (Theorem II.3.9).

Proof. Define

$$H_1 := \{x \in H : \ \|T(t)x\| = \|T^*(t)x\| = \|x\| \quad \text{for all } t \geq 0\}.$$

Observe that for every $0 \neq x \in H_1$ and $t \geq 0$,

$$\|x\|^2 = \langle T(t)x, T(t)x \rangle = \langle T^*(t)T(t)x, x \rangle \leq \|T^*(t)T(t)x\| \|x\| \leq \|x\|^2.$$

Therefore, by the equality in the Cauchy-Schwarz inequality and the positivity of $\|x\|^2$, we obtain $T^*(t)T(t)x = x$. Analogously, $T(t)T^*(t)x = x$. On the other hand, every x with these two properties belongs to H_1. So we proved the equality

$$H_1 = \{x \in H : \ T^*(t)T(t)x = T(t)T^*(t)x = x \quad \text{for all } t \geq 0\} \qquad \text{(III.27)}$$

which shows, in particular, that H_1 is the maximal (closed) subspace on which $T(\cdot)$ is unitary. The $T(t)$- and $T^*(t)$-invariance of H_1 follows from the definition of H_1 and the equality $T^*(t)T(t) = T(t)T^*(t)$ on H_1.

To show (b) take $x \in H_2 := H_1^\perp$. We first note that H_2 is $T(\cdot)$- and $T^*(\cdot)$-invariant since H_1 is so. Suppose now that $T(t)x$ does not converge weakly to zero as $t \to \infty$, or, equivalently, that there exists $y \in H$, $\varepsilon > 0$ and a sequence $\{t_n\}_{n=1}^\infty$ such that $|\langle T(t_n)x, y \rangle| \geq \varepsilon$ for every $n \in \mathbb{N}$. On the other hand, there exists a weakly converging subsequence of $\{T(t_n)x\}_{n=1}^\infty$. For convenience we denote the subsequence again by $\{t_n\}_{n=1}^\infty$ and its limit by x_0. The closedness and $T(t)$-invariance of H_2 imply that $x_0 \in H_2$.

For a fixed $t_0 \geq 0$ we obtain

$$\|T^*(t_0)T(t_0)T(t)x - T(t)x\|^2$$
$$= \|T^*(t_0)T(t + t_0)x\|^2 - 2\langle T^*(t_0)T(t + t_0)x, T(t)x \rangle + \|T(t)x\|^2$$
$$\leq \|T(t + t_0)x\|^2 - 2\|T(t + t_0)x\|^2 + \|T(t)x\|^2$$
$$= \|T(t)x\|^2 - \|T(t + t_0)x\|^2.$$

The right-hand side converges to zero as $t \to \infty$ since the function $t \mapsto \|T(t)x\|$ is monotone decreasing on $\mathbb{R}_+$. Therefore we obtain $\lim_{t \to \infty} \|T^*(t_0)T(t_0)T(t)x - T(t)x\| = 0$.

We now remember that weak-$\lim_{n\to\infty} T(t_n)x = x_0$. This implies immediately that weak-$\lim_{n\to\infty} T^*(t_0)T(t_0)T(t_n)x = T^*(t_0)T(t_0)x_0$. By the considerations above, we have on the other hand that weak-$\lim_{n\to\infty} T^*(t_0)T(t_0)T(t_n)x = x_0$ and therefore $T^*(t_0)T(t_0)x_0 = x_0$. One shows analogously that $T(t_0)T^*(t_0)x_0 = x_0$ and $x_0 \in H_1$. By $H_1 \cup H_2 = \{0\}$ this implies $x_0 = 0$, which is a contradiction.

Analogously one shows that the restriction of $T^*(\cdot)$ to H_2 converges weakly to zero as well. □

Remark 4.8. The restriction of $T(\cdot)$ to the subspace H_2 in Theorem 4.7 is *completely non-unitary* (c.n.u. for short), i.e., there is no subspace of H_2 on which the restriction of $T(\cdot)$ becomes unitary. In other words, Theorem 4.7 states that every Hilbert space contraction semigroup can be decomposed into a unitary and a c.n.u. part and the c.n.u. part is weakly stable.

For a systematic study of completely non-unitary semigroups as well as an alternative proof of Theorem 4.7 (except weak stability on H_2) using unitary dilation theory see the monograph of Sz.-Nagy and Foiaş [238, Proposition 9.8.3].

On the other hand, the following theorem gives a decomposition into weakly stable and weakly unstable parts due to Foguel [86]. We give a simplified proof of it.

Theorem 4.9 (Foguel). *Let $T(\cdot)$ be a contraction semigroup on a Hilbert space H. Define*

$$W := \{x \in H : \lim_{t\to 0}\langle T(t)x, x\rangle = 0\}.$$

Then

$$W = \{x \in H : \lim_{t\to 0} T(t)x = 0 \text{ weakly}\} = \{x \in H : \lim_{t\to 0} T^*(t)x = 0 \text{ weakly}\},$$

W is a closed $T(\cdot)$- and $T^*(\cdot)$-invariant subspace of H and the restriction of $T(\cdot)$ to $W^\perp$ is unitary.

Proof. We first show that weak-$\lim_{t\to\infty} T(t)x = 0$ for a fixed $x \in W$. By Theorem 4.7 we may assume that $x \in H_1$. If we take $S := \overline{\text{lin}}\{T(t)x : t \geq 0\}$, then by the decomposition $H = S \oplus S^\perp$ it is enough to show that $\lim_{t\to\infty}\langle T(t)x, y\rangle = 0$ for all $y \in S$. For $y := T(t_0)x$ we obtain

$$\langle T(t)x, y\rangle = \langle T^*(t_0)T(t)x, x\rangle = \langle T(t - t_0)x, y\rangle \to 0 \quad \text{for } t_0 \leq t \to \infty,$$

since the restriction of $T(\cdot)$ to H_1 is unitary. The density of $\text{lin}\{T(t)x : t \geq 0\}$ in S implies $\lim_{t\to\infty}\langle T(t)x, y\rangle = 0$ for every $y \in S$ and therefore weak-$\lim_{t\to\infty} T(t)x = 0$. Analogously, weak-$\lim_{t\to\infty} T^*(t)x = 0$. The converse implication, the closedness and the invariance of W are clear.

The last assertion of the theorem follows directly from Theorem 4.7. □

Combining Theorem 4.7 and Theorem 4.9 we obtain the following decomposition into three orthogonal subspaces.

Theorem 4.10. *Let $T(\cdot)$ be a contraction semigroup on a Hilbert space H. Then H is the orthogonal sum of three closed $T(\cdot)$- and $T^*(\cdot)$-invariant subspaces H_1, H_2 and H_3 such that the restrictions $T_1(\cdot)$, $T_2(\cdot)$ and $T_3(\cdot)$ satisfy the following.*

(a) *$T_1(\cdot)$ is unitary and has no non-zero weakly stable orbit;*

(b) *$T_2(\cdot)$ is unitary and weakly stable;*

(c) *$T_3(\cdot)$ is completely non-unitary and weakly stable.*

As in the discrete case (see Subsection II.3.2), we see from the above theorem that a characterisation of weak stability for unitary groups on Hilbert spaces is of special importance. We will discuss more aspects of this problem in Subsection IV.1.1.

At the end of this subsection we present the following classical result describing the part $T_3(\cdot)$ in the above theorem if the semigroup consists of isometries, see Foiaş, Sz.-Nagy [238, Theorem III.9.3].

Theorem 4.11 (Wold decomposition). *Let $T(\cdot)$ be an isometric C_0-semigroup on a Hilbert space H. Then H can be decomposed into an orthogonal sum $H = H_0 \oplus H_1$ of $T(\cdot)$-invariant subspaces such that the restriction of $T(\cdot)$ to H_0 is a unitary semigroup and the restriction of $T(\cdot)$ to H_1 is unitarily equivalent to the unilateral shift on $L^2(\mathbb{R}_+, Y)$ for a Hilbert space Y. In addition, $\dim Y = \dim(\mathrm{rg}\, V)^\perp$ where V is the cogenerator of $T(\cdot)$.*

4.3 Characterisation via resolvent

In this subsection we discuss a resolvent approach due to Chill, Tomilov [49], see also Eisner, Farkas, Nagel, Serény [67].

The following main result gives some sufficient conditions for weak stability, see Chill, Tomilov [49] and Eisner, Farkas, Nagel, Sereny [67].

Theorem 4.12. *Let $T(\cdot)$ be a C_0-semigroup on a Banach space X with generator A satisfying $s_0(A) \leq 0$. For $x \in X$ and $y \in X'$ fixed, consider the following assertions.*

(a) $$\int_0^1 \int_{-\infty}^\infty |\langle R^2(a+is, A)x, y\rangle|\, ds\, da < \infty;$$

(b) $$\lim_{a \to 0+} a \int_{-\infty}^\infty |\langle R^2(a+is, A)x, y\rangle|\, ds = 0;$$

(c) $$\lim_{t \to \infty} \langle T(t)x, y\rangle = 0.$$

Then (a)$\Rightarrow$(b)$\Rightarrow$(c). In particular, if $T(\cdot)$ is bounded and (a) or (b) holds for all x from a dense subset of X and all y from a dense subset of X', then $(T(t))_{t\geq 0}$ is weakly stable.

Proof. We first show that (a) implies (b).

From the theory of Hardy spaces, see, e.g., Rosenblum, Rovnyak [223, Theorem 5.20], we know that the function $f : (0, 1) \mapsto \mathbb{R}_+$ defined by

$$f(a) := \int_{-\infty}^{\infty} |\langle R^2(a + is, A)x, y\rangle| \, ds$$

is monotone decreasing for $a > 0$. Assume now that (b) is not true. Then there exists a monotone decreasing null sequence $\{a_n\}_{n=1}^{\infty}$ such that

$$a_n f(a_n) \geq c \tag{III.28}$$

holds for some $c > 0$ and all $n \in N$.

Take now $n, m \in \mathbb{N}$ such that $a_n \leq \frac{a_m}{2}$. By (III.28) and the monotonicity of f we have

$$\int_{a_n}^{a_m} f(a) \, da \geq f(a_m)(a_m - a_n) \geq \frac{c}{a_m}(a_m - a_n) = c\left(1 - \frac{a_n}{a_m}\right) \geq \frac{c}{2}.$$

This contradicts (a) and the implication (a)$\Rightarrow$(b) is proved.

It remains to show that (b) implies (c). By (b) we have

$$\int_{-\infty}^{\infty} |\langle R^2(a + is, A)x, y\rangle| \, ds < \infty$$

for every $a > 0$. Since $s_0(A) \leq 0$, Theorem I.3.9 (see also Theorem I.3.8) implies the inverse Laplace transform representation

$$\langle T(t)x, y\rangle = \frac{1}{2\pi t} \int_{-\infty}^{\infty} e^{(a+is)t} \langle R^2(a + is, A)x, y\rangle \, ds \tag{III.29}$$

for all $a > 0$. We now take $t = \frac{1}{a}$ to obtain

$$|\langle T(t)x, y\rangle| \leq a \int_{-\infty}^{\infty} |\langle R^2(a + is, A)x, y\rangle| \, ds \to 0$$

as $a \to 0+$, so $t = \frac{1}{a} \to \infty$, proving (c).

The last part of the theorem follows from the standard density argument. $\square$

Remark 4.13. Convergence of the integrals in (b) and hence in (a) in Theorem 4.12 for all $x \in X$ and $y \in X'$ implies $s_0(A) = \omega_0(T)$ by Corollary 2.19, and hence is not necessary for weak stability of a C_0-semigroup on a Banach space.

A useful necessary and sufficient resolvent condition for weak stability is still unknown. In particular, it is not clear whether condition (b) in Theorem b holds for all x and y from dense subsets for weakly stable C_0-semigroups on Banach spaces (and even for unitary groups on Hilbert spaces).

5   Almost weak stability

In this section we investigate almost weak stability, a concept looking more complicated, but being much easier to characterise than weak stability. We will follow Eisner, Farkas, Nagel, Serény [67].

5.1   Characterisation

In this part we always assume $T(\cdot)$ to be a relatively compact semigroup. (This is for example the case if $T(\cdot)$ is bounded and X is reflexive, see Example I.1.7(a)). Since every weakly stable semigroup is relatively weakly compact, it is not a severe restriction.

We begin with a list of equivalent properties motivating our definition of almost weak stability.

Theorem 5.1. *Let $(T(t))_{t\geq 0}$ be a relatively weakly compact C_0-semigroup on a Banach space X with generator A. The following assertions are equivalent.*

(i) $0 \in \overline{\{T(t)x : t \geq 0\}}^{\sigma}$ *for every $x \in X$;*

(i') $0 \in \overline{\{T(t) : t \geq 0\}}^{\mathcal{L}_\sigma}$;

(ii)   *For every $x \in X$ there exists a sequence $\{t_j\}_{j=1}^{\infty} \subset \mathbb{R}_+$ converging to ∞ such that $\lim_{j\to\infty} T(t_j)x = 0$ weakly;*

(iii)   *For every $x \in X$ there exists a set $M \subset \mathbb{R}_+$ with density 1 such that*
$$\lim_{t\to\infty,\, t\in M} T(t)x = 0 \text{ weakly;}$$

(iv) $\displaystyle\lim_{t\to\infty} \frac{1}{t} \int_0^t |\langle T(s)x, y\rangle|\, ds = 0$ *for all $x \in X$, $y \in X'$;*

(v) $\displaystyle\lim_{a\to 0+} a \int_{-\infty}^{\infty} |\langle R(a+is, A)x, y\rangle|^2\, ds = 0$ *for all $x \in X$, $y \in X'$;*

(vi) $\displaystyle\lim_{a\to 0+} aR(a+is, A)x = 0$ *for all $x \in X$ and $s \in \mathbb{R}$;*

(vii) $P_\sigma(A) \cap i\mathbb{R} = \emptyset$, *i.e., A has no eigenvalues on the imaginary axis.*

If, in addition, X' is separable, then the conditions above are also equivalent to

(ii*)   *There exists a sequence $\{t_j\}_{j=1}^{\infty} \subset R_+$ converging to ∞ such that $\lim_{j\to\infty} T(t_j) = 0$ in the weak operator topology;*

(iii*)   *There exists a set $M \subset \mathbb{R}_+$ with density 1 such that $\lim_{t\to\infty,\, t\in M} T(t) = 0$ in the weak operator topology.*

Analogously to the discrete case, the (asymptotic) density of a measurable set $M \subset \mathbb{R}_+$ is
$$d(M) := \lim_{t\to\infty} \frac{1}{t}\lambda([0,t] \cap M),$$

whenever the limit exists (here λ is the Lebesgue measure on $\mathbb{R}$). Note that 1 is the greatest possible density.

We will use the following elementary fact, being the continuous analogue of Lemma II.4.2.

Lemma 5.2 (Koopman–von Neumann, 1932). *Let $f : \mathbb{R}_+ \to \mathbb{R}_+$ be measurable and bounded. The following assertions are equivalent.*

(a) $\displaystyle \lim_{t\to\infty} \frac{1}{t} \int_0^t f(s) \, ds = 0;$

(b) *There exists a set $M \subset \mathbb{R}_+$ with density 1 such that $\displaystyle \lim_{t\to\infty, \, t\in M} f(t) = 0.$*

Proof. Assume (b). Then we have for $C := \sup_{t\geq 0} f(t)$ that

$$\frac{1}{t}\int_0^t f(s)\,ds = \frac{1}{t}\int_0^t f(s)\mathbf{1}_M(s)\,ds + \frac{1}{t}\int_0^t f(s)\mathbf{1}_{M^c}(s)\,ds$$
$$\leq \frac{1}{t}\int_0^t f(s)\mathbf{1}_M(s)\,ds + \frac{C}{t}\lambda(M^c \cap [0,t]),$$

where M^c denotes the complement of M in $\mathbb{R}_+$. The first summand on the right-hand side converges to 0 as $t \to \infty$ since convergence implies Cesàro convergence, and the second summand converges to 0 as $t \to \infty$ since M^c has density 0, proving (a).

For the converse implication assume (a) and define for $k \in \mathbb{N}$ the sets

$$N_k := \left\{ s \in \mathbb{R}_+ : \ f(s) > \frac{1}{k} \right\}.$$

We have $N_k \subset N_{k+1}$ for all $k \in \mathbb{N}$. Moreover, since $\mathbf{1}_{N_k}(s) < kf(s)$ holds for every $s \geq 0$, (a) implies $d(N_k) = 0$. Therefore there exists an increasing sequence $\{t_k\}_{k=1}^\infty$ with $\lim_{k\to\infty} t_k = \infty$ such that

$$\frac{1}{t}\int_0^t \mathbf{1}_{N_k}(s)\,ds < \frac{1}{k} \quad \text{for all } t \geq t_{k-1}. \tag{III.30}$$

We show that the set

$$M := \bigcup_{k=1}^\infty N_k^c \cap [t_{k-1}, t_k]$$

has the desired properties.

Observe first that for $t \in [t_{k-1}, t_k] \cap M$ we have $t \notin N_k$, i.e., $f(t) \leq \frac{1}{k}$, which implies

$$\lim_{t\to\infty, \, t\in M} f(t) = 0.$$

Moreover, since the sets N_k^c descrease in k, (III.30) implies that for every k and $t \in [t_{k-1}, t_k]$

$$\frac{1}{t} \int_0^t 1_M(s)\, ds \geq \frac{1}{t} \int_0^t 1_{N_k^c}(s)\, ds = 1 - \frac{1}{t} \int_0^t 1_{N_k}(s)\, ds > 1 - \frac{1}{k}.$$

So $d(M) = 1$ and (b) is proved. $\qquad\qquad\qquad\qquad\qquad\qquad\qquad\qquad\qquad\square$

Proof of Theorem 5.1. The proof of the implication (i′) $\Rightarrow$ (i) is trivial. The implication (i) $\Rightarrow$ (ii) holds since in Banach spaces weak compactness and weak sequential compactness coincide by the Eberlein–Šmulian theorem (Theorem I.1.1).

If (vii) does not hold, then (ii) cannot be true by the spectral mapping theorem for the point spectrum (see Engel, Nagel [78, Theorem IV.3.7]), hence (ii) $\Rightarrow$ (vii).

The implication (vii) $\Rightarrow$ (i′) is the main consequence of the Jacobs–Glicksberg–de Leeuw decomposition (Theorem I.1.19) and follows from the construction in its proof, see Engel, Nagel [78], p. 313.

This proves the equivalences (i) $\Leftrightarrow$ (i′) $\Leftrightarrow$ (ii) $\Leftrightarrow$ (vii).

(vi) $\Leftrightarrow$ (vii): Since the semigroup $(T(t))_{t\geq 0}$ is bounded and mean ergodic by Theorem I.2.25, we have by Proposition I.2.24 that the decomposition $X = \ker A \oplus \overline{\mathrm{rg}\, A}$ holds and the limit

$$Px := \lim_{a \to 0+} aR(a, A)x$$

exists for all $x \in X$ with a projection P onto $\ker A$. Therefore, $0 \notin P\sigma(A)$ if and only if $P = 0$. Take now $s \in \mathbb{R}$. The semigroup $(e^{ist}T(t))_{t\geq 0}$ is also relatively weakly compact and hence mean ergodic. Repeating the argument for this semigroup we obtain (vi) $\Leftrightarrow$ (vii).

(i′) $\Rightarrow$ (iii): Let $S := \overline{\{T(t) : t \geq 0\}}^{\mathcal{L}_\sigma} \subseteq \mathcal{L}(X)$ which is a compact semitopological semigroup if considered with the usual multiplication and the weak operator topology. By (i) we have $0 \in S$. Define the operators $\tilde{T}(t) : C(S) \to C(S)$ by

$$(\tilde{T}(t)f)(R) := f(T(t)R), \quad f \in C(S), \ R \in S.$$

By Nagel (ed.) [196], Lemma B-II.3.2, $(\tilde{T}(t))_{t\geq 0}$ is a C_0-semigroup on $C(S)$.

By Example I.1.7 (c), the set $\{f(T(t)\cdot) : t \geq 0\}$ is relatively weakly compact in $C(S)$ for every $f \in C(S)$. It means that every orbit $\{\tilde{T}(t)f : t \geq 0\}$ is relatively weakly compact, and, by Lemma I.1.6, $(\tilde{T}(t))_{t\geq 0}$ is a relatively weakly compact semigroup.

Denote by $\tilde{P}$ the mean ergodic projection of $(\tilde{T}(t))_{t\geq 0}$. We have

$$\mathrm{Fix}(\tilde{T}) = \bigcap_{t\geq 0} \mathrm{Fix}(\tilde{T}(t)) = \langle 1 \rangle.$$

Indeed, for $f \in \mathrm{Fix}(\tilde{T})$ one has $f(T(t)I) = f(I)$ for all $t \geq 0$ and therefore f must be constant. Hence $\tilde{P}f$ is constant for every $f \in C(S)$. By definition of the ergodic projection

$$(\tilde{P}f)(0) = \lim_{t \to \infty} \frac{1}{t} \int_0^t \tilde{T}(s)f(0)\,ds = f(0). \tag{III.31}$$

Thus we have

$$(\tilde{P}f)(R) = f(0) \cdot 1, \qquad f \in C(S),\ R \in S. \tag{III.32}$$

Take now $x \in X$. By Theorem I.1.5 and its proof (see Dunford, Schwartz [63, p. 434]), the weak topology on the orbit $\{T(t)x : t \geq 0\}$ is metrisable and coincides with the topology induced by some sequence $\{y_n\}_{n=1}^\infty \subset X' \setminus \{0\}$. Consider $f_{x,n} \in C(S)$ defined by

$$f_{x,n}(R) := |\langle Rx, \tfrac{y_n}{\|y_n\|} \rangle|, \qquad R \in S,$$

and $f_x \in C(S)$ defined by

$$f_x(R) := \sum_{n \in \mathbb{N}} \frac{1}{2^n} f_{x,n}(R), \qquad R \in S.$$

By (III.32) we obtain

$$0 = \lim_{t \to \infty} \frac{1}{t} \int_0^t \tilde{T}(s)f_{x,y}(I)\,ds = \lim_{t \to \infty} \frac{1}{t} \int_0^t f_x(T(s))\,ds.$$

Lemma 5.2 applied to the continuous and bounded function $\mathbb{R}_+ \ni t \mapsto f(T(t)I)$ yields a set $M \subset \mathbb{R}$ with density 1 such that

$$\lim_{t \to \infty,\, t \in M} f_x(T(t)) = 0.$$

By definition of f_x and by the fact that the weak topology on the orbit is induced by $\{y_n\}_{n=1}^\infty$ we have in particular that

$$\lim_{t \to \infty,\, t \in M} T(t)x = 0 \quad \text{weakly,}$$

proving (iii).

(iii) $\Rightarrow$ (iv) follows directly from Lemma 5.2.

(iv) $\Rightarrow$ (vii) holds by the spectral mapping theorem for the point spectrum, see Engel, Nagel [78, Theorem IV.3.7].

(iv) $\Leftrightarrow$ (v): Clearly, the semigroup $(T(t))_{t \geq 0}$ is bounded. Take $x \in X$, $y \in X'$ and let $a > 0$. By the Plancherel theorem applied to the function $t \mapsto e^{-at}\langle T(t)x, y \rangle$ we have

$$\int_{-\infty}^\infty |\langle R(a + is, A)x, y \rangle|^2\,ds = 2\pi \int_0^\infty e^{-2at} |\langle T(t)x, y \rangle|^2\,dt.$$

We obtain by the equivalence of Abel and Cesàro limits, see Lemma I.2.23,

$$\lim_{a\to 0+} a \int_{-\infty}^{\infty} |\langle R(a+is, A)x, y\rangle|^2 \, ds = \lim_{a\to 0+} a \int_{0}^{\infty} e^{-2at} |\langle T(s)x, y\rangle|^2 \, ds$$

$$= \lim_{t\to\infty} \frac{1}{t} \int_{0}^{t} |\langle T(s)x, y\rangle|^2 \, ds. \quad \text{(III.33)}$$

Note that for a bounded continuous function $f : \mathbb{R}_+ \to \mathbb{R}_+$ with $C := \sup f(\mathbb{R}_+)$ we have

$$\left(\frac{1}{Ct} \int_{0}^{t} f^2(s) \, ds\right)^2 \leq \left(\frac{1}{t} \int_{0}^{t} f(s) \, ds\right)^2 \leq \frac{1}{t} \int_{0}^{t} f^2(s) \, ds,$$

which together with (III.33) gives the equivalence of (iv) and (v).

For the remaining part of the theorem suppose X' to be separable. Then so is X, and we can take dense subsets $\{x_n \neq 0 : n \in \mathbb{N}\} \subseteq X$ and $\{y_m \neq 0 : m \in \mathbb{N}\} \subseteq X'$. Consider the functions

$$f_{n,m} : S \to \mathbb{R}, \qquad f_{n,m}(R) := \left|\left\langle R\frac{x_n}{\|x_n\|}, \frac{y_m}{\|y_m\|}\right\rangle\right|, \quad n, m \in \mathbb{N},$$

which are continuous and uniformly bounded in $n, m \in \mathbb{N}$. Define the function

$$f : S \to \mathbb{R}, \qquad f(R) := \sum_{n,m\in\mathbb{N}} \frac{1}{2^{n+m}} f_{n,m}(R).$$

Then clearly $f \in C(S)$. Thus, as in the proof of the implication (i') $\Rightarrow$ (iii), i.e., using (III.31) we obtain

$$\lim_{t\to\infty} \frac{1}{t} \int_{0}^{t} f(T(s)I) \, ds = 0.$$

Hence, applying Lemma 5.2 to the continuous and bounded function $\mathbb{R}_+ \ni t \mapsto f(T(t)I)$, we obtain the existence of a set M with density 1 such that $f(T(t)) \to 0$ as $t \to \infty$, $t \in M$. In particular, $|\langle T(t)x_n, y_m\rangle| \to 0$ for all $n, m \in \mathbb{N}$ as $t \to \infty$, $t \in M$, which, together with the boundedness of $(T(t))_{t\geq 0}$, proves the implication (i') $\Rightarrow$ (iii*). The implications (iii*) $\Rightarrow$ (ii*) $\Rightarrow$ (i') are straightforward, hence the proof is complete. $\qquad\square$

The above theorem shows that starting from

"no eigenvalues of the generator on the imaginary axis",

one arrives at properties like (iii) on the orbits of the semigroup. This justifies the following terminology.

Definition 5.3. A relatively weakly compact C_0-semigroup is *almost weakly stable* if it satisfies condition (iii) in Theorem 5.1.

Historical remark 5.4. Theorem 5.1 and especially the implication (vii) $\Rightarrow$ (iii) were first proved for discrete semigroups and has a long history, see Remark II.4.4. The conditions (i), (iii) and (iv) were studied by Hiai [129] even for strongly measurable semigroups. He related it to the discrete case as well. See also Kühne [156, 157]. The implication (vii)$\Rightarrow$(i) appears also in Ruess, Summers [224] in a more abstract context. Note that the equivalence (vii)$\Leftrightarrow$(iv) for contraction semigroups on Hilbert spaces is a consequence of the Wiener theorem, see Goldstein [101].

Remark 5.5. The conditions in Theorem 5.1 are of quite different natures. Conditions (i)–(iv) as well as (ii*) and (iii*) give information on the behaviour of the semigroup, while conditions (v)–(vii) deal with the spectrum and the resolvent of the generator near the imaginary axis. Among them condition (vii) apparently is the simplest to verify. Moreover, as in the discrete case one can add the equivalent condition

(iv$'$) $\lim\limits_{t\to\infty} \sup\limits_{y\in X', \|y\|\leq 1} \dfrac{1}{t}\int_0^t |\langle T(s)x, y\rangle| = 0$ for every $x \in X$,

see Hiai [129].

The equivalence (i$'$) $\Leftrightarrow$ (v) in Theorem 5.1 is a weak analogue to Tomilov's characterisation of strong stability given in Corollary 3.23.

Remark 5.6. It is also quite surprising that the set $M \subset \mathbb{R}_+$ of density 1 appearing in condition (iii) has certain algebraic structure. As we saw in Theorem 4.4, if M contains a relatively dense set, then the semigroup $T(\cdot)$ is automatically weakly stable. In other words, if we have an almost weakly but not weakly stable semigroup, then the set M in condition (iii) does not contain any relatively dense set.

We now can state the following version of the Jacobs–Glicksberg–de Leeuw decomposition (see Theorem I.1.19).

Theorem 5.7 (Jacobs–Glicksberg–de Leeuw decomposition, extended version). *Let* X *be a Banach space and* $T(\cdot)$ *be a relatively weakly compact* C_0-*semigroup on* X. *Then* $X = X_r \oplus X_s$, *where*

$X_r := \overline{\lin}\{x \in X : T(t)x = e^{i\alpha t}x \text{ for some } \alpha \in \mathbb{R} \text{ and all } t \geq 0\}$,

$X_s := \{x \in X : \lim\limits_{t\to\infty,\ t\in M} T(t)x = 0 \text{ weakly for some set } M \subset \mathbb{R}_+ \text{ with density 1}\}$.

Jan van Neerven (private communication) pointed out that there is a version of Theorem 5.1 for single orbits.

Corollary 5.8. *Let* A *generate a* C_0-*semigroup* $T(\cdot)$ *on a Banach space* X *and* $x \in X$. *Assume that the orbit* $\{T(t)x : t \geq 0\}$ *is relatively weakly compact in* X *and that the restriction of* $T(\cdot)$ *to* $\overline{\lin}\{T(t)x : t \geq 0\}$ *is bounded. Then there is a holomorphic continuation of the resolvent function* $R(\cdot, T)x$ *to* $\{\lambda : Re(\lambda) > 0\}$ *denoted by* $R_x(\cdot)$ *and the following assertions are equivalent.*

(i) $0 \in \overline{\{T(t)x : t \geq 0\}}^{\sigma}$.

(ii) *There exists a sequence $\{t_j\}_{j=1}^{\infty}$ converging to ∞ such that $\lim_{j\to\infty} T(t_j)x = 0$ weakly.*

(iii) *There exists a set $M \subset \mathbb{R}_+$ with density 1 such that $\lim_{t\to\infty,\, t\in M} T(t)x = 0$ weakly.*

(iv) $\lim_{t\to\infty} \dfrac{1}{t} \displaystyle\int_0^t |\langle T(s)x, y\rangle|\, ds = 0$ *for all $y \in X'$.*

(v) $\lim_{a\to 0+} a \displaystyle\int_{-\infty}^{\infty} |\langle R_x(a + is), y\rangle|^2\, ds = 0$ *for all $y \in X'$.*

(vi) $\lim_{a\to 0+} a R_x(a + is) = 0$ *for all $s \in \mathbb{R}$.*

(vii) *The restriction of A to $\overline{\operatorname{lin}}\{T(t)x : t \geq 0\}$ has no eigenvalue on the imaginary axis.*

Proof. For the first part of the theorem define

$$R_x(\lambda) := \int_0^{\infty} e^{-\lambda t} T(t)x\, dt \quad \text{whenever } \operatorname{Re} \lambda > 1.$$

Denote now by Z the closed linear span of the orbit $\{T(t)x : t \geq 0\}$. Then Z is a $T(\cdot)$-invariant closed subspace of X and we can take a restriction of $T(\cdot)$ denoted by $T_Z(\cdot)$ which is, by Lemma I.1.6, relatively weakly compact as well. By the uniqueness of the Laplace transform we obtain that $R(A_Z, \lambda)x = R_x(\lambda)$ for every λ with $\operatorname{Re}(\lambda) > 0$, where A_Z denotes the generator of $T_Z(\cdot)$.

The rest follows from the canonical decomposition $X' = Z' \oplus Z^0$ with $Z^0 := \{y \in X' : \langle z, y\rangle = 0 \text{ for all } z \in Z\}$ and Theorem 5.1. $\qquad\square$

5.2 Concrete example

We now give a concrete example of an almost weakly stable but not weakly stable C_0-semigroup, again following Eisner, Farkas, Nagel, Serény [67].

Example 5.9. As in Nagel (ed.) [196], p. 206, we start from a flow on $\mathbb{C}\backslash\{0\}$ with the following properties:

1) The orbits starting in z with $|z| \neq 1$ spiral towards the unit circle Γ;

2) 1 is the fixed point of φ and $\Gamma \backslash \{1\}$ is a homoclinic orbit, i.e., $\lim_{t\to-\infty} \varphi_t(z) = \lim_{t\to\infty} \varphi_t(z) = 1$ for every $z \in \Gamma$.

A concrete example comes from the differential equation in polar coordinates $(r, \omega) = (r(t), \omega(t))$:

$$\begin{cases} \dot{r} &= 1 - r, \\ \dot{\omega} &= 1 + (r^2 - 2r\cos\omega), \end{cases}$$

see the following picture.

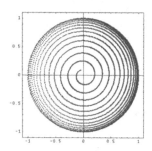

Take $x_0 \in \mathbb{C}$ with $0 < |x_0| < 1$ and denote by $S_{x_0} := \{\varphi_t(x_0) : t \geq 0\}$ the orbit starting from x_0. Then $S := S_{x_0} \cup \Gamma$ is compact for the usual topology of $\mathbb{C}$.

We define a multiplication on S as follows. For $x = \varphi_t(x_0)$ and $y = \varphi_s(x_0)$ we put

$$xy := \varphi_{t+s}(x_0).$$

For $x \in \Gamma$, $x = \lim_{n \to \infty} x_n$, $x_n = \varphi_{t_n}(x_0) \in S_{x_0}$ and $y = \varphi_s(x_0) \in S_{x_0}$, we define $xy = yx := \lim_{n \to \infty} x_n y$. Note that by $|x_n y - \varphi_s(x)| = |\varphi_s(x_n) - \varphi_s(x)| \leq C|x_n - x| \xrightarrow[n \to \infty]{} 0$ the definition is correct and satisfies

$$xy = \varphi_s(x).$$

For $x, y \in \Gamma$ we define $xy := 1$. This multiplication on S is separately continuous and makes S a semi-topological semigroup (see Engel, Nagel [78], Sec. V.2).

Consider now the Banach space $X := C(S)$. By Example I.1.7 (c) the set

$$\{f(s \cdot) : s \in S\} \subset C(S)$$

is relatively weakly compact for every $f \in C(S)$. By definition of the multiplication on S this implies that

$$\{f(\varphi_t(\cdot)) : t \geq 0\}$$

is relatively weakly compact in $C(S)$. Consider the semigroup induced by the flow, i.e.,

$$(T(t)f)(x) := f(\varphi_t(x)), \quad f \in C(S), \ x \in S.$$

By the above, each orbit $\{T(t)f : t \geq 0\}$ is relatively weakly compact in $C(S)$ and hence, by Lemma I.1.6, $(T(t))_{t \geq 0}$ is weakly compact. Note that the strong continuity of $(T(t))_{t \geq 0}$ follows, as shown in Nagel (ed.) [196, Lemma B-II.3.2], from the separate continuity of the flow.

Next, we take $X_0 := \{f \in C(S) : f(1) = 0\}$ and identify it with the Banach lattice $C_0(S \setminus \{1\})$. Then both subspaces in the decomposition $C(S) = X_0 \oplus \langle 1 \rangle$ are invariant under $(T(t))_{t \geq 0}$. Denote by $(T_0(t))_{t \geq 0}$ the semigroup restricted to X_0 and by A_0 its generator. The semigroup $(T_0(t))_{t \geq 0}$ remains relatively weakly compact.

Since $\mathrm{Fix}(T_0) = \bigcap_{t \geq 0} \mathrm{Fix}(T_0(t)) = \{0\}$, we have that $0 \notin P_\sigma(A_0)$. Moreover, $P_\sigma(A_0) \cap i\mathbb{R} = \emptyset$ holds, which implies by Theorem 5.1 that $(T_0(t))_{t \geq 0}$ is almost weakly stable.

To see that $(T_0(t))_{t \geq 0}$ is not weakly stable it is enough to consider $\delta_{x_0} \in X_0'$. Since

$$\langle T_0(t)f, \delta_{x_0} \rangle = f(\varphi(t, x_0)), \ f \in X_0,$$

$f(\Gamma)$ always belongs to the closure of $\{\langle T_0(t)f, \delta_{x_0} \rangle : \ t \geq 0\}$ and hence the semigroup $(T_0(t))_{t \geq 0}$ cannot be weakly stable.

We summarise the above as follows.

Theorem 5.10. *There exist a locally compact space Ω and a positive, relatively weakly compact C_0-semigroup on $C_0(\Omega)$ which is almost weakly but not weakly stable.*

The above phenomenon cannot occur for positive semigroups on special Banach lattices; for details and discussion see Chill, Tomilov [50].

Theorem 5.11 (Groh, Neubrander [112, Theorem. 3.2]; Chill, Tomilov [50, Theorem. 7.7]). *For a bounded, positive, mean ergodic C_0-semigroup $(T(t))_{t \geq 0}$ on a Banach lattice X with generator $(A, D(A))$, the following assertions hold.*

(i) *If $X \cong L^1(\Omega, \mu)$, then $P\sigma(A') \cap i\mathbb{R} = \emptyset$ is equivalent to the strong stability of $(T(t))_{t \geq 0}$.*

(ii) *If $X \cong C(K)$, K compact, then $P\sigma(A') \cap i\mathbb{R} = \emptyset$ is equivalent to the uniform exponential stability of $(T(t))_{t \geq 0}$.*

Example 5.9 above shows that (ii) does not hold in spaces $C_0(\Omega)$, Ω locally compact.

6   Category theorems

In this section we compare weak and almost weak stability and show that (at least on Hilbert spaces) these two related notions differ drastically. Analogously to the discrete case (see Section II.5), we show following Eisner, Serény [70] that a "typical" unitary C_0-group as well as a "typical" isometric C_0-semigroup on a separable Hilbert space is almost weakly but not weakly stable for an appropriate topology. An even stronger result will be shown in Section IV.3.

6.1 Unitary case

We start from the set $\mathcal{U}^{\text{cont}}$ of all unitary C_0-groups on a separable infinite-dimensional Hilbert space H.

A first step in our construction is the following density result.

Proposition 6.1. *For every $n \in \mathbb{N}$ the set of all periodic unitary C_0-groups with period greater than n is dense in $\mathcal{U}^{\text{cont}}$ in the norm topology uniform on compact time intervals.*

Proof. Take $U(\cdot) \in \mathcal{U}^{\text{cont}}$ and $n \in \mathbb{N}$. By the spectral theorem H is isomorphic to $L^2(\Omega, \mu)$ for some locally compact space Ω and finite measure μ and $U(\cdot)$ is unitarily equivalent to a multiplication semigroup $\tilde{U}(\cdot)$ with

$$(\tilde{U}(t)f)(\omega) = e^{itq(\omega)} f(\omega) \quad \forall \omega \in \Omega, \ t \geq 0, \ f \in L^2(\Omega, \mu)$$

for some measurable $q : \Omega \to \mathbb{R}$.

We approximate the semigroup $\tilde{U}(\cdot)$ as follows. For $k > n$ define

$$q_k(\omega) := \frac{2\pi j}{k} \quad \forall \omega \in q^{-1}\left(\left[\frac{2\pi j}{k}, \frac{2\pi(j+1)}{k}\right]\right), \ j \in \mathbb{Z}.$$

The multiplication operator with $e^{itq_k(\cdot)}$ we denote by $\tilde{V}_k(t)$, and $\tilde{V}_k(\cdot)$ is a periodic unitary group with period greater than or equal to $k > n$. Moreover,

$$\|\tilde{U}(t)f - \tilde{V}_k(t)f\| = \int_{\Omega} |e^{itq(\omega)} - e^{itq_k(\omega)}|^2 \|f(\omega)\|^2 \, d\omega$$

$$\leq 2|t| \sup_{\omega} |q(\omega) - q_k(\omega)| \cdot \|f\|^2 = \frac{4\pi|t|}{k} \|f\|^2$$

holds. So $\lim_{k \to \infty} \|\tilde{U}(t) - \tilde{V}_k(t)\| = 0$ uniformly for t in compact intervals, and the proposition is proved. $\qquad\square$

Remark 6.2. By a modification of the proof of Proposition 6.1 one can choose the approximating periodic unitary groups $\tilde{V}(\cdot)$ to have bounded generators.

For the second step we need the following lemma.

Lemma 6.3. *Let H be a separable infinite-dimensional Hilbert space. Then there exists a sequence $\{U_n(\cdot)\}_{n=1}^{\infty}$ of almost weakly stable unitary groups with bounded generator satisfying $\lim_{n \to \infty} \|U_n(t) - I\| = 0$ uniformly in t in compact intervals.*

Proof. By isomorphy of all separable infinite-dimensional Hilbert spaces we can assume that $H = L^2(\mathbb{R})$ with respect to the Lebesgue measure.

Take $n \in \mathbb{N}$ and define $U_n(\cdot)$ on $L^2(\mathbb{R})$ by

$$(U_n(t)f)(s) := e^{\frac{itq(s)}{n}} f(s), \quad s \in \mathbb{R}, \ f \in L^2(\mathbb{R}),$$

where $q : \mathbb{R} \to [0, 1]$ is a strictly monotone increasing function.

Then all $U_n(\cdot)$ are almost weakly stable by Theorem 5.1 and we have

$$\|U_n(t) - I\| = \sup_{s \in \mathbb{R}} \left| e^{\frac{it q(s)}{n}} - 1 \right| \leq [\text{for } t \leq \pi n] \leq \left| e^{\frac{it}{n}} - 1 \right| \leq \frac{2t}{n} \to 0, \quad n \to \infty,$$

uniformly for t in compact intervals.   □

We consider the topology on the space $\mathcal{U}^{\mathrm{cont}}$ coming from the metric

$$d(U(\cdot), V(\cdot)) := \sum_{n,j=1}^{\infty} \frac{\sup_{t \in [-n,n]} \|U(t)x_j - V(t)x_j\|}{2^j \|x_j\|} \quad \text{for } U, V \in \mathcal{U}^{\mathrm{cont}},$$

where $\{x_j\}_{j=1}^{\infty}$ is some fixed dense subset of $H \setminus \{0\}$. Note that this topology corresponds to the strong convergence uniform on compact time intervals in $\mathbb{R}$ and is a continuous analogue of the strong*-topology for operators used in Subsection II.5.1.

We further denote by $\mathcal{S}_{\mathcal{U}}^{\mathrm{cont}}$ the set of all weakly stable unitary groups on H and by $\mathcal{W}_{\mathcal{U}}^{\mathrm{cont}}$ the set all almost weakly stable unitary groups on H.

The following result shows density of $\mathcal{W}_{\mathcal{U}}^{\mathrm{cont}}$ in $\mathcal{U}^{\mathrm{cont}}$.

Proposition 6.4. *The set $\mathcal{W}_{\mathcal{U}}^{\mathrm{cont}}$ of all almost weakly stable unitary groups with bounded generator is dense in $\mathcal{U}^{\mathrm{cont}}$.*

Proof. By Proposition 6.1 it is enough to approximate periodic unitary groups by almost weakly unitary groups. Let $U(\cdot)$ be a periodic unitary group with period τ. Take $\varepsilon > 0$, $n \in \mathbb{N}$, $x_1, \ldots, x_n \in H \setminus \{0\}$ and $t_0 > 0$. We have to find an almost weakly stable unitary group $V(\cdot)$ with $\|U(t)x_j - V(t)x_j\| \leq \varepsilon$ for all $j = 1, \ldots, n$ and all t with $|t| \leq t_0$.

By Engel, Nagel [78, Theorem IV.2.26] we have

$$H = \overline{\oplus^{\perp}_{k \in \mathbb{Z}} \ker \left(A - \frac{2\pi i k}{\tau} \right)},$$

where A denotes the generator of $U(\cdot)$. So we can assume without loss of generality that $\{x_j\}_{j=1}^{n}$ is an orthonormal system of eigenvalues of A.

Define now the $T(\cdot)$-invariant subspace $H_0 := \lim\{x_1, \ldots, x_n\}$ and $B := A$ on H_0. Further, since H is separable, the decomposition

$$H = \oplus^{\perp}_{k \in \mathbb{N}} H_k$$

holds, where $\dim H_k = \dim H_0$ for every $k \in \mathbb{N}$. For a fixed orthonormal basis $\{e_j^k\}_{j=1}^{n}$ of each H_k we define $B e_j^k := B x_j$ and extend B to a bounded linear operator on H.

From the construction follows that

$$H = \ker \left(B - \frac{2\pi i \lambda_1}{\tau} \right) \oplus^{\perp} \ldots \oplus^{\perp} \ker \left(B - \frac{2\pi i \lambda_n}{\tau} \right),$$

where $\frac{2\pi i \lambda_j}{\tau}$ is the eigenvalue of A (and therefore of B) corresponding to the eigenvector x_j.

Let $X_j := \ker\left(B - \frac{2\pi i \lambda_1}{\tau}\right)$ for every $j = 1, \ldots, n$. On every X_j the operator B is equal to $\frac{2\pi i \lambda_j}{\tau} I$. Note further that all X_j are infinite-dimensional. By Lemma 6.3 for every j there exists an almost weakly stable unitary group $T_j(\cdot)$ on X_j such that $\left\| T_j(t) - e^{\frac{2\pi t i \lambda_j}{\tau}} I \right\| < \varepsilon$ for every t with $|t| \leq t_0$. Denote now by $T(\cdot)$ the orthogonal sum of $T_j(\cdot)$ which is a weakly stable unitary group with bounded generator. Moreover, we obtain that

$$\| T(t)x_j - U(t)x_j \| = \left\| T(t)x_j - e^{\frac{2\pi i t \lambda_j}{\tau}} x_j \right\| \leq \varepsilon$$

for every t with $|t| \leq t_0$ and the proposition is proved. $\qquad\square$

We now prove a category theorem for weakly and almost weakly stable unitary groups, being analogous to its discrete counterpart given in Subsection II.5.1.

Theorem 6.5. *The set $\mathcal{S}_{\mathcal{U}}^{cont}$ of weakly stable unitary groups is of first category and the set $\mathcal{W}_{\mathcal{U}}^{cont}$ of almost weakly stable unitary groups is residual in $\mathcal{U}^{cont}$.*

Proof. We first prove that $\mathcal{S}_{\mathcal{U}}^{cont}$ is of first category in $\mathcal{U}^{cont}$. Fix $x \in H$ with $\|x\| = 1$ and consider

$$M_t := \left\{ U(\cdot) \in \mathcal{U}^{cont} : |\langle U(t)x, x \rangle| \leq \frac{1}{2} \right\}.$$

Note that all sets M_t are closed.

For every weakly stable $U(\cdot) \in \mathcal{U}^{cont}$ there exists $t > 0$ such that $U \in M_s$ for all $s \geq t$, i.e., $U(\cdot) \in N_t := \cap_{s \geq t} M_t$. So we obtain

$$\mathcal{S}_{\mathcal{U}}^{cont} \subset \bigcup_{t>0} N_t. \tag{III.34}$$

Since all N_t are closed, it remains to show that $\mathcal{U}^{cont} \setminus N_t$ is dense for every t.

Fix $t > 0$ and let $U(\cdot)$ be a periodic unitary group. Then $U(\cdot) \notin M_s$ for some $s \geq t$ and therefore $U(\cdot) \notin N_t$. Since, by Proposition 6.1, periodic unitary groups are dense in $\mathcal{U}^{cont}$, the set $\mathcal{S}_{\mathcal{U}}^{cont}$ is of first category.

To show that $\mathcal{W}_{\mathcal{U}}^{cont}$ is residual we take a dense subspace $D = \{x_j\}_{j=1}^{\infty}$ of H and define

$$W_{jkt} := \left\{ U(\cdot) \in \mathcal{U}^{cont} : |\langle U(t)x_j, x_j \rangle| < \frac{1}{k} \right\}.$$

All these sets are open, and therefore the sets $W_{jk} := \cup_{t>0} W_{jkt}$ are also open.

We now show the equality

$$\mathcal{W}_{\mathcal{U}}^{cont} = \bigcap_{j,k=1}^{\infty} W_{jk}. \tag{III.35}$$

The inclusion "⊂" follows from the definition of almost weak stability. To prove the converse inclusion we take $U(\cdot) \notin \mathcal{W}_{\mathcal{U}}^{\mathrm{cont}}$ and $t > 0$. Then there exists $x \in H$ with $\|x\| = 1$ and $\varphi \in \mathbb{R}$ such that $U(t)x = e^{it\varphi}x$ for all $t > 0$, which implies $|\langle U(t)x, x\rangle| = 1$. Take $x_j \in D$ with $\|x_j - x\| \leq \frac{1}{4}$. Then we have

$$|\langle U(t)x_j, x_j\rangle| = |\langle U(t)(x - x_j), x - x_j\rangle + \langle U(t)x, x\rangle - \langle U(t)x, x - x_j\rangle$$
$$- \langle U(t)(x - x_j), x\rangle| \geq 1 - \|x - x_j\|^2 - 2\|x - x_j\| > \frac{1}{3}.$$

So $U(\cdot) \notin W_{j3}$, hence $U(\cdot) \notin \cap_{j,k=1}^{\infty} W_{jk}$, and equality (III.35) holds. Therefore $\mathcal{W}_{\mathcal{U}}^{\mathrm{cont}}$ is residual as a dense countable intersection of open sets. □

6.2 Isometric case

In this subsection we consider the space $\mathcal{I}^{\mathrm{cont}}$ of all isometric C_0-semigroups on H endowed with the strong topology uniform on compact time intervals and prove category results as in the previous subsection. We again assume H to be separable and infinite-dimensional. Note that $\mathcal{I}^{\mathrm{cont}}$ is a complete metric space with respect to the metric given by the formula

$$d(T(\cdot), S(\cdot)) := \sum_{n,j=1}^{\infty} \frac{\sup_{t \in [0,n]} \|T(t)x_j - S(t)x_j\|}{2^j \|x_j\|} \quad \text{for } T(\cdot), S(\cdot) \in \mathcal{I}^{\mathrm{cont}},$$

where $\{x_j\}_{j=1}^{\infty}$ is a fixed dense subset of $H \setminus \{0\}$.

We further denote by $\mathcal{S}_{\mathcal{I}}^{\mathrm{cont}}$ the set of all weakly stable, and by $\mathcal{W}_{\mathcal{I}}^{\mathrm{cont}}$ the set of all almost weakly stable isometric C_0-semigroups on H.

The main tool is the classical Wold decomposition from Theorem 4.11. In addition, we need the following easy lemma (see also Peller [211]).

Lemma 6.6. *Let Y be a Hilbert space and let $R(\cdot)$ be the right shift semigroup on $H := l^2(\mathbb{N}, Y)$. Then there exists a sequence $\{U_n(\cdot)\}_{n=1}^{\infty}$ of periodic unitary groups on H converging strongly to $R(\cdot)$ uniformly on compact time intervals.*

Proof. For every $n \in \mathbb{N}$ we define $U_n(t)$ by

$$(U_n(t)f)(s) := \begin{cases} f(s), & s \geq n; \\ R_n(t)f(s), & s \in [0,n], \end{cases}$$

where $R_n(\cdot)$ denotes the n-periodic right shift on the space $L^2([0,n], Y)$. Then every $U_n(\cdot)$ is a C_0-semigroup on $L^2(\mathbb{R}_+, Y)$ which is isometric and n-periodic, and therefore unitary.

Fix $f \in L^2(\mathbb{R}_+, Y)$ and $T > 0$. Then for $t \leq T$ and $n > T$ we have

$$\|U_n(t)f - R(t)f\|^2 = \int_n^\infty \|f(s) - f(s+n-t)\|^2 ds + \int_0^t \|f(s+n-t)\|^2 ds$$

$$\leq \int_n^\infty \|f(s)\|^2 ds + \int_{n-t}^\infty \|f(s)\|^2 ds + \int_{n-t}^n \|f(s)\|^2 ds$$

$$= 2 \int_{n-t}^\infty \|f(s)\|^2 ds \leq 2 \int_{n-T}^\infty \|f(s)\|^2 ds \xrightarrow[n\to\infty]{} 0$$

uniformly on $[0,T]$, and the lemma is proved. $\qquad\square$

As a consequence we obtain the following density result for periodic C_0-groups in $\mathcal{I}^{cont}$.

Proposition 6.7. *The set of all periodic unitary C_0-groups is dense in $\mathcal{I}^{cont}$.*

Proof. Let $V(\cdot)$ be an isometric semigroup on H. Then by Theorem 4.11 the orthogonal decomposition $H = H_0 \oplus H_1$ holds, where the restriction $V_0(\cdot)$ of $V(\cdot)$ to H_0 is unitary, H_1 is unitarily equivalent to $L^2(\mathbb{R}_+, Y)$ for some Y and the restriction $V_1(\cdot)$ of $V(\cdot)$ on H_1 corresponds to the right shift semigroup on $L^2(\mathbb{R}_+, Y)$. By Proposition 6.1 and Lemma 6.6 and their proofs we can approximate both semigroups $V_0(\cdot)$ and $V_1(\cdot)$ by unitary periodic ones with period in $\mathbb{N}$ and the assertion follows. $\qquad\square$

Also the following density result for almost weakly stable semigroups is a consequence of Wold's decomposition and the results of the previous subsection.

Proposition 6.8. *The set $\mathcal{W}_\mathcal{I}^{cont}$ of almost weakly stable isometric C_0-semigroups is dense in $\mathcal{I}^{cont}$.*

Proof. Let $V(\cdot)$ be an isometric C_0-semigroup on H, H_0 and H_1 the orthogonal subspaces from Theorem 4.11 and $V_0(\cdot)$, $V_1(\cdot)$ the corresponding restrictions of $V(\cdot)$. By Lemma 6.6, the semigroup $V_1(\cdot)$ can be approximated by unitary C_0-groups on H_1. The assertion now follows from Proposition 6.4. $\qquad\square$

Using the same idea as in the proofs of Theorem 6.5, Propositions 6.7 and 6.8, we obtain the following category theorem for weakly and almost weakly stable isometric C_0-semigroups.

Theorem 6.9. *The set $\mathcal{S}_\mathcal{I}^{cont}$ of all weakly stable isometric C_0-semigroups is of first category and the set $\mathcal{W}_\mathcal{I}^{cont}$ of all almost weakly stable isometric C_0-semigroups is residual in $\mathcal{I}^{cont}$.*

6.3 Contractive case: remarks

It is not clear whether one can prove results analogous to Theorems 6.5 and 6.9 for contractive C_0-semigroups (as in the discrete case in Section II.5.3). We point out the main difficulty.

Let $\mathcal{C}^{\text{cont}}$ denote the set of all contraction semigroups on H endowed with the metric

$$d(T(\cdot), S(\cdot)) := \sum_{n,i,j=1}^{\infty} \frac{\sup_{t \in [0,n]} |\langle T(t)x_i, x_j \rangle - \langle S(t)x_i, x_j \rangle|}{2^{i+j+n} \|x_i\| \|x_j\|} \quad \text{for } T, S \in \mathcal{C}^{\text{cont}},$$

where $\{x_j\}_{j=1}^{\infty}$ is a fixed dense subset of $H \setminus \{0\}$. The corresponding convergence is the weak convergence of semigroups uniform on compact time intervals. Note that this metric is a continuous analogue of the metric used in Section II.5.3.

The following example, see Eisner, Serény [71], shows that the space $\mathcal{C}^{\text{cont}}$ is *not* complete (or compact) in general, and hence the Baire category theorem cannot be applied as it was done in the proofs of Theorems 6.5 and 6.9.

Example 6.10. Consider $X := l^p$, $1 \leq p < \infty$, and the operators $\tilde{A}_n$ defined by

$$\tilde{A}_n(x_1, \ldots, x_n, x_{n+1}, \ldots, x_{2n}, \ldots)$$
$$:= (x_{n+1}, x_{n+2}, \ldots, x_{2n}, x_1, x_2, \ldots, x_n, x_{2n+1}, x_{2n+2}, \ldots)$$

exchanging the first n coordinates of a vector with its next n coordinates. Then $\|\tilde{A}_n\| \leq 1$ implies that $\tilde{A}_n$ generates a C_0-semigroup $\tilde{T}_n(\cdot)$ satisfying $\|\tilde{T}_n(t)\| \leq e^t$ for every $n \in \mathbb{N}$ and $t \geq 0$.

The operators $\tilde{A}_n$ converge weakly to zero as $n \to \infty$. Moreover, $\tilde{A}_n^2 = I$ for every $n \in \mathbb{N}$. Therefore

$$\tilde{T}_n(t) = \sum_{k=0}^{\infty} \frac{t^k \tilde{A}_n^k}{k!} = \sum_{k=0}^{\infty} \frac{t^{2k+1}}{(2k+1)!} \tilde{A}_n + \sum_{k=0}^{\infty} \frac{t^{2k}}{(2k)!} I$$
$$= \frac{e^t - e^{-t}}{2} \tilde{A}_n + \frac{e^t + e^{-t}}{2} I \xrightarrow{\sigma} \frac{e^t + e^{-t}}{2} I \neq I,$$

where "σ" denotes the weak operator topology, and this convergence is uniform on compact time intervals. However, the limit does not satisfy the semigroup law.

By rescaling $A_n := \tilde{A}_n - I$ we obtain a sequence of contractive semigroups converging weakly and uniformly on compact time intervals to a family which is not a semigroup while the bounded generators converge weakly to $-I$ (which is a generator).

Thus it is not clear whether $\mathcal{C}^{\text{cont}}$ is a Baire space.

Note finally that related residuality results appear in the context of positive semigroups, see Bartoszek and Kuna [17] for recent category results for Markov semigroups on the Schatten class C_1 and Lasota, Myjak [160] for an analogous result for stochastic semigroups.

Remark 6.11. In Section IV.3 we improve the above results using the notion of rigidity.

7 Stability via Lyapunov's equation

In the last section of this chapter we characterise stability and boundedness of C_0-semigroups on Hilbert spaces via a Lyapunov equation. This goes back to Lyapunov [166] in 1892 where he investigated uniform exponential stability of matrix semigroups.

As in the discrete case, we use a positivity approach as, e.g., Nagel (ed.) [196, Section D-IV.2], Nagel, Rhandi [199], Groh, Neubrander [112], Batty, Robinson [28] and Alber [4, 5]. A different approach is in Guo, Zwart [113], see also Datko [55]. In what follows, only the characterisation of uniform exponential stability is classical, while the other results are recent. Our main tools are implemented semigroups as defined in Subsection I.4.4.

7.1 Uniform exponential stability

We start with the infinite-dimensional version of Lyapunov's result and characterise uniform exponential stability.

Theorem 7.1. *Let $T(\cdot)$ be a C_0-semigroup on a Hilbert space H with generator A and $\mathcal{T}(\cdot)$ be the corresponding implemented semigroup with generator $\mathcal{A}$. Then the following assertions are equivalent.*

(i) *$T(\cdot)$ is uniformly exponentially stable.*

(ii) *$\mathcal{T}(\cdot)$ is uniformly exponentially stable.*

(iii) *$0 \in \rho(\mathcal{A})$ and $0 \le R(0, \mathcal{A})$.*

(iv) *There exists $0 \le Q \in \mathcal{L}(H)$ with $Q(D(A)) \subset D(A^*)$ such that*

$$\mathcal{A}Q = A^*Q + QA = -I \text{ on } D(A). \tag{III.36}$$

(v) *There exists $0 \le Q \in \mathcal{L}(H)$ with $Q(D(A)) \subset D(A^*)$ such that*

$$\mathcal{A}Q = A^*Q + QA \le -I \text{ on } D(A).$$

In this case, the solution Q of (III.36) is unique and given as $Q = R(0, \mathcal{A})I = \int_0^\infty T(t)^*T(t)\,dt$, where the integral is defined with respect to the strong operator topology.

Equation (III.36) is called *Lyapunov equation*.

Proof. (i)⇔(ii) follows directly from Lemma I.4.8(a), and (ii)⇔(iii) by Theorem I.4.9.

We now show (iii)⇒(iv). By Theorem I.4.9, (iii) implies that $\mathcal{T}(\cdot)$ is uniformly exponentially stable and the representation $R(0, \mathcal{A})Xx = \int_0^\infty \mathcal{T}(t)Xx\,dt$ holds for every $X \in \mathcal{L}(H)$ and $x \in H$. Define $Q \in \mathcal{L}(H)$ by

$$Q := R(0, \mathcal{A})I = -\mathcal{A}^{-1}I \in D(\mathcal{A})$$

which is positive semidefinite and satisfies

$$Qx = R(0, A)Ix = \int_0^\infty T(t)^* T(t) x \, dt$$

for every $x \in H$. Moreover, we have by the definition of $\mathcal{A}$ that $A^* Q + QA = AQ = -I$, and (iv) follows.

(iv)$\Rightarrow$(v) is trivial, and it remains to show that (v)$\Rightarrow$(i).

Assume (v). We have by (v)

$$\int_0^t \mathcal{T}(s) I \, ds \leq - \int_0^t \mathcal{T}(s) \mathcal{A} Q \, ds = Q - \mathcal{T}(t) Q \leq Q,$$

where all integrals are defined strongly in $\mathcal{L}(H)$. Since the family $\left(\int_0^t \mathcal{T}(s) I \, ds \right)_{t \geq 0}$ is monotone increasing in $\mathcal{L}(H)$, the integral $\int_0^\infty \mathcal{T}(s) I \, ds$ converges for the weak operator topology, cf. Lemma I.4.1(b). In particular, the integral

$$\int_0^\infty \langle \mathcal{T}(s) I x, x \rangle \, ds = \int_0^\infty \| T(s) x \|^2 \, ds$$

converges for every $x \in H$, and therefore $T(\cdot)$ is uniformly exponentially stable by Datko's theorem, see Theorem 2.10. $\square$

7.2 Strong stability

The idea to characterise also strong stability via Lyapunov equations appears in Guo, Zwart [113]. We give an alternative proof.

Theorem 7.2. *Let $T(\cdot)$ be a C_0-semigroup with generator A on a Hilbert space H, and let $\mathcal{T}(\cdot)$ be its implemented semigroup with generator $\mathcal{A}$. Then the following assertions are equivalent.*

(i) *$T(\cdot)$ is strongly stable.*

(ii) *$\lim_{t \to \infty} \langle \mathcal{T}(t) Sx, x \rangle = 0$ for all $S \in \mathcal{L}(H)$ and $x \in H$.*

(iii) *For every $a > 0$ there exist (unique) $0 \leq Q_a, \tilde{Q}_a \in \mathcal{L}(H)$ with $Q_a(D(A)) \subset D(A^*)$ and $\tilde{Q}_a(D(A^*)) \subset D(A)$ satisfying*

$$(A - aI)^* Q_a + Q_a(A - aI) = -I \quad \text{on } D(A),$$
$$(A - aI)\tilde{Q}_a + \tilde{Q}_a(A - aI)^* = -I \quad \text{on } D(A^*)$$

such that

$$\lim_{a \to 0+} a \langle Q_a x, x \rangle = 0 \quad \text{and} \quad \sup_{a > 0} a \langle \tilde{Q}_a x, x \rangle < \infty \quad \text{for all } x \in H. \quad \text{(III.37)}$$

In this case,

$$Q_a = R(a, \mathcal{A})I = \int_0^\infty e^{-at} T(t)^* T(t) \, dt, \qquad \text{(III.38)}$$

$$\tilde{Q}_a = R(a, \mathcal{A}^*)I = \int_0^\infty e^{-at} T(t) T(t)^* \, dt, \qquad \text{(III.39)}$$

where the integrals are defined strongly.

Proof. (ii)$\Rightarrow$(i) follows directly from $\langle T(t)Ix, x \rangle = \|T(t)x\|^2$.

(i)$\Rightarrow$(ii): Let $x \in H$ and $S \in \mathcal{L}(H)_{sa}$. Then there is a constant $c > 0$ such that $-cI \leq S \leq cI$. This implies $|\langle T(t)Sx, x \rangle| \leq c|\langle T(t)Ix, x \rangle| = c\|T(t)x\|^2$. By (i) we thus have $\lim_{n\to\infty} \langle T(t)Sx, x \rangle = 0$ for every $S \in \mathcal{L}(H)_{sa}$ and hence for every $S \in \mathcal{L}(H)$.

We prove (i)$\Leftrightarrow$(iii) simultaneously. Assume that (i) or (iii) holds. From Theorem 7.1 applied to the semigroups $(e^{-at}T(t))_{t\geq 0}$ we know that the rescaled Lyapunov equations in (iii) are satisfied by (the unique) $Q_a := R(a, \mathcal{A})I = \int_0^\infty e^{-at} T^*(t)T(t) \, dt$, where the integral is defined strongly on H. Therefore we have

$$\langle Q_a x, x \rangle = \int_0^\infty e^{-at} \|T(t)x\|^2 \, dt.$$

Analogously, $\tilde{Q}_a := R(a, \mathcal{A}^*)I$ and

$$\langle \tilde{Q}_a x, x \rangle = \int_0^\infty e^{-at} \|T(t)^* x\|^2 \, dt.$$

The condition $\tilde{Q}_a(D(A^*)) \subset D(A)$ follows from $D(A^{**}) = D(A)$ and Theorem 7.1 (iv). By Lemma I.2.6 we have that (III.37) is equivalent to

$$\lim_{t\to\infty} \frac{1}{t} \int_0^t \|T(s)x\|^2 \, ds = 0 \quad \text{and} \quad \sup_{t>0} \frac{1}{t} \int_0^t \|T(s)^* x\|^2 \, ds < \infty$$

for all $x \in H$. By Theorem 3.7 for $p = q = 2$ this is equivalent to strong stability of $T(\cdot)$, and the proof is finished. $\qquad\qquad\square$

Remark 7.3. By the above proof, one can replace "$=$" by "$\leq$" in the rescaled Lyapunov equations in (iii).

7.3 Boundedness

We now characterise boundedness of C_0-semigroups on Hilbert spaces via Lyapunov equations, see Guo, Zwart [113] for the result and a different proof.

Theorem 7.4. *Let $T(\cdot)$ be a C_0-semigroup on a Hilbert space H. Then the following assertions are equivalent.*

(i) $T(\cdot)$ *is bounded.*

(ii) *For every $a > 0$ there exist (unique) $0 \leq Q_a, \tilde{Q}_a \in \mathcal{L}(H)$ with $Q_a(D(A)) \subset D(A^*)$ and $\tilde{Q}_a(D(A^*)) \subset D(A)$ such that*

$$(A - aI)^* Q_a + Q_a(A - aI) = -I \text{ on } D(A),$$
$$(A - aI)\tilde{Q}_a + \tilde{Q}_a(A - aI)^* = -I \text{ on } D(A^*)$$

with

$$\sup_{a>0} a\langle Q_a x, x \rangle < \infty \quad and \quad \sup_{a>0} a\langle \tilde{Q}_a x, x \rangle < \infty \quad \forall x \in H. \tag{III.40}$$

In this case, Q_a and $\tilde{Q}_a$ satisfy representations (III.38) and (III.39), respectively, where the integrals are defined strongly.

The proof is analogous to the one of Theorem 7.2 using Theorem 1.6 instead of Theorem 3.7. Moreover, one can replace "$= -I$" by "$\leq -I$" in both equalities in (ii) as well. Moreover, one can replace (III.40) by the condition used in Guo, Zwart [113]

$$\sup_{a>0} a\|Q_a\| < \infty \quad and \quad \sup_{a>0} a\|\tilde{Q}_a\| < \infty$$

which is equivalent to (III.40) by the polarisation identity and the uniform boundedness principle.

Final remark

As in the discrete case, see Section II.6, weak stability is much more delicate to treat. The main problem occurs again when $T(\cdot)$ is a unitary group.

Open question 7.5. Is it possible to characterise weak stability of C_0-semigroups on Hilbert spaces via some kind of Lyapunov equations?

Chapter IV

Connections to ergodic and measure theory

Ergodic theory has motivated important notions and results in operator theory such as, e.g., the mean ergodic theorem. In this chapter we connect stability of operators and C_0-semigroups back to analogous notions in harmonic analysis and ergodic theory. In the last section we describe "typical" asymptotics of discrete and continuous semigroups using rigidity, a notion well-known in these two areas.

1 Stability and the Rajchman property

In this section we show parallels between operator theory and measure theory. It turns out that weak stability of operators and C_0-semigroups has a direct analogue in measure theory called the Rajchman property.

Rajchman measures are certain measures on the unit circle or on the real line (or, more generally, on a locally compact abelian group). Starting his investigation in 1922, Rajchman was motivated by questions on the uniqueness of trigonometric series. Since then many people have worked on Rajchman measures (see, e.g., Rajchman [216, 217], Milicer-Grużewska [185], Lyons [174, 175, 176, 177, 178], Bluhm [34], Koerner [151], Goldstein [98, 99], Ransford [218]), and we refer to Lyons [179] for a historical overview.

1.1 From stability to the Rajchman property and back

We now introduce the Rajchman property for measures and relate it to weak stability of unitary operators and semigroups.

Discrete case

We first consider an example which will help us to understand the general situation.

Example 1.1. Let μ be some probability measure on the unit circle Γ and consider the multiplication operator T on the space $H := L^2(\Gamma, \mu)$ defined by

$$(Tf)(z) := zf(z), \quad z \in \Gamma.$$

Then T is unitary, hence not strongly stable, and we are interested in its weak stability.

Since $z \in P_\sigma(T)$ if and only if $\mu(\{z\}) > 0$, the Jacobs–Glicksberg–de Leeuw decomposition (Theorem I.1.15) implies the following equivalence.

T is almost weakly stable $\iff$ μ is a continuous measure, i.e.,

$$\mu(\{z\}) = 0 \text{ for all } z \in \Gamma.$$

To characterise weak stability, we first observe that

$$\langle T^n f, f \rangle = \int_0^{2\pi} e^{ins} |f(e^{is})|^2 \, d\mu(s)$$

for every $f \in H$, where we identify Γ with $[0, 2\pi]$. Therefore, if T is weakly stable and for $f = 1$, we obtain that the Fourier coefficients $\hat{\mu}_n$ of μ must satisfy

$$\hat{\mu}_n := \int_0^{2\pi} e^{ins} \, d\mu(s) \to 0 \quad \text{as } n \to \infty. \tag{IV.1}$$

Conversely, assume that condition (IV.1) holds. Take f, g defined by $f(s) := e^{ims}$ and $g(s) := e^{ils}$ and observe that

$$\langle T^n f, g \rangle = \int_0^{2\pi} e^{i(n+m-l)s} \, d\mu(s) \to 0 \quad \text{as } n \to \infty.$$

By the standard linearity and density argument we obtain $\lim_{n\to\infty} \langle T^n f, g \rangle = 0$ for every $f, g \in L^2(\Gamma, \mu)$, i.e., T is weakly stable.

Note that a unitary operator is weakly stable if and only if its inverse is, hence in condition (IV.1) we can also take $n \to -\infty$.

This and Theorem II.4.1 prove the following proposition (see Lyons [175] for the first equivalence in (a)).

Proposition 1.2. *For the operator T defined above, the following holds.*

(a) *T is weakly stable $\iff \lim_{n\to\infty} \hat{\mu}_n = 0 \iff \lim_{|n|\to\infty} \hat{\mu}_n = 0$.*

(b) *T is almost weakly stable $\iff \mu$ is continuous $\iff \lim_{j\to\infty} \hat{\mu}_{n_j} = 0$ for some $n_j \to \infty \iff \lim_{j\to\infty} \hat{\mu}_{n_j} = 0$ for some $\{n_j\}_{j=1}^\infty$ with density 1.*

Remark 1.3. The fact that a measure μ is continuous if and only if its Fourier coefficients converge to zero on a subsequence of $\mathbb{N}$ (or, equivalently, of $\mathbb{Z}$) with density 1 also follows from Wiener's lemma, see e.g. Katznelson [146, p. 42], and Lemma II.4.2.

Measures satisfying (a) in the above proposition were given their own name.

Definition 1.4. A Radon measure on Γ is called *Rajchman* if its Fourier coefficients converge to zero.

Proposition 1.2 states that the operator T is weakly stable if and only if the measure μ is Rajchman.

For the properties of Rajchman measures we refer to Lyons [175, 179]. We only remark that every absolutely continuous measure is Rajchman by the Riemann-Lebesgue lemma and every Rajchman measure is continuous by Wiener's lemma, but none of the converse implications holds. The first example of a singular Rajchman measure was a modified Cantor-Lebesgue measure constructed by Menshov [184] in 1916. Not absolutely continuous Rajchman measures with remarkable properties have been constructed later, see, e.g., Bluhm [34] and Körner [151]. On the other side, the classical Cantor-Lebesgue middle-third measure is an example of a continuous measure which is not Rajchman.

By our considerations above, each continuous non-Rajchman measure μ on Γ induces an almost weakly but not weakly stable unitary operator on $L^p(\Gamma, \mu)$.

By the spectral theorem (see e.g. Conway [52, Theorem IX.4.6]), one can reduce the general situation to the previous example.

Indeed, consider a contraction T on a Hilbert space H. By Theorem III.4.7, its restriction T_1 to the subspace $H_1 := \{x : \|T^n x\| = \|T^{*n} x\| = \|x\| \ \forall n \in \mathbb{N}\}$ is unitary and its restriction to $H_1^{\perp}$ is weakly stable. So T is weakly stable if and only if T_1 is weakly stable.

We now apply the spectral theorem to the unitary operator T_1 and obtain, for each $x \in H_1$, a measure μ_x on Γ such that the restriction of T to $\overline{\mathrm{lin}}\{T^n x : n = 0, 1, 2, \ldots\}$ is isomorphic to the multiplication operator $M_z f(z) := z f(z)$ on $L^2(\Gamma, \mu_x)$. So we are in the context of Example 1.1 and see that

$$\lim_{n \to \infty} T^n x = 0 \text{ weakly} \quad \Longleftrightarrow \quad \mu_x \text{ is Rajchman.}$$

Thus we have the following characterisation of stability of unitary operators via spectral measures.

Proposition 1.5. *Let H be a Hilbert space with orthonormal basis S and U be a unitary operator on H. Then U is weakly stable if and only if the spectral measures μ_x are Rajchman for every $x \in S$. Moreover, U is almost weakly stable if and only if μ_x is continuous for every $x \in S$.*

This gives a measure theoretic approach to weak stability of operators. However, since there is no simple characterisation of Rajchman measures, this approach is difficult to use in concrete situations.

Continuous case

We now relate weak stability of C_0-semigroups to the Rajchman property of measures on the real line. The results are analogous to the discrete case discussed in the previous subsection.

We again begin with an example which is typical for the general situation.

Example 1.6. Let μ be a probability measure on $\mathbb{R}$ and, on the space $H :=$ $L^2(\mathbb{R}, \mu)$, take the multiplication operator

$$(Af)(s) := isf(s), \quad s \in \mathbb{R},$$

with its maximal domain $D(A) := \{f \in H : g \in H \text{ for } g(s) := isf(s)\}$. The C_0-group $T(\cdot)$ generated by A is

$$(T(t)f)(s) := e^{ist}f(s), \quad s, t \in \mathbb{R}, \ f \in H.$$

It is unitary and hence not strongly stable, but we can ask for weak or almost weak stability.

Note that $is \in P_\sigma(A)$ if and only if $\mu(\{s\}) > 0$ and the Jacobs–Glicksberg–de Leeuw decomposition (Theorem I.1.19) implies

$$T(\cdot) \text{ is almost weakly stable} \iff \mu \text{ is continuous, i.e., } \mu(\{s\}) = 0 \ \forall s \in \mathbb{R}.$$

On the other hand, we see that

$$\langle T(t)f, f \rangle = \int_{-\infty}^{\infty} e^{ist} |f(s)|^2 d\mu(s)$$

holds for every $f \in H$. Therefore, if $T(\cdot)$ is weakly stable, then

$$\mathcal{F}\mu(t) := \int_{-\infty}^{\infty} e^{ist} d\mu(s) \to 0 \quad \text{as } t \to \infty, \tag{IV.2}$$

where $\mathcal{F}\mu$ denotes the Fourier transform of μ.

Conversely, if (IV.2) holds, then $\lim_{t\to\infty} \langle T(t)f, f \rangle = 0$ for every f having constant absolute value. Since the linear span of $\{e^{in\cdot}\}_{n=-\infty}^{\infty}$ is dense in H and $T(\cdot)$ is contractive, $\lim_{t\to\infty} \langle T(t)f, f \rangle = 0$ for every $f \in H$, so by the polarisation identity $T(\cdot)$ is weakly stable. Note further that a unitary group is weakly stable for $t \to +\infty$ if and only if it is weakly stable for $t \to -\infty$. This proves the following proposition (see Lyons [175]).

Proposition 1.7. *The following assertions hold for the above semigroup $T(\cdot)$.*

(a) $T(\cdot)$ *is weakly stable* $\iff \lim_{t\to\infty} \mathcal{F}\mu(t) = 0 \iff \lim_{|t|\to\infty} \mathcal{F}\mu(t) = 0$.

(b) $T(\cdot)$ *is almost weakly stable* $\iff \mu$ *is continuous* $\iff \lim_{j\to\infty} \mathcal{F}\mu(t_j) = 0$ *for some* $|t_j| \to \infty \iff \lim_{|t|\to\infty, \ t\in M} \mathcal{F}\mu(t) = 0$ *for some* $M \subset \mathbb{R}$ *with density* 1.

As before, a finite measure on $\mathbb{R}$ is called *Rajchman* if its Fourier transform converges to zero at infinity.

We refer to Lyons [175, 179] and Goldstein [98, 99] for a brief overview on Rajchman measures on the real line and their properties (the second author uses the term "Riemann–Lebesgue measures"). We just mention that, as in the discrete case, absolutely continuous measures are always Rajchman by the Riemann–Lebesgue lemma, and all Rajchman measures are continuous by Wiener's theorem. However, there are continuous measures which are not Rajchman and Rajchman measures which are not absolutely continuous (see Lyons [179] and Goldstein [99]).

It is now a consequence of the considerations above that each continuous non-Rajchman measure gives rise to an almost weakly but not weakly stable unitary C_0-group. For a concrete example of a unitary group with bounded generator whose spectral measures are not Rajchman see Engel, Nagel [78, p. 316].

Finally, we can connect Rajchman measures on $\mathbb{R}$ to Rajchman measures on Γ (as defined in Example 1.1). Indeed, for a probability measure μ on $\mathbb{R}$ take its image ν under the map $s \mapsto e^{is}$. Then μ is Rajchman if and only if ν is so.

Using the spectral theorem for unitary operators on Hilbert spaces, questions concerning weak (and almost weak) stability of C_0-semigroups can be reduced to the previous example.

Indeed, consider an arbitrary contraction semigroup $T(\cdot)$ on a Hilbert space H. By Theorem III.4.7 the restriction $T_1(\cdot)$ of $T(\cdot)$ to the subspace $W := \{x : \|T(t)x\| = \|T^*(t)x\| = \|x\| \ \forall t \geq 0\}$ is unitary and the restriction to $W^\perp$ is weakly stable. In order to check weak stability, it remains to investigate the unitary (semi)group $T_1(\cdot)$.

Applying the spectral theorem to A_1 we obtain for each $x \in H_1$ a measure μ_x on $\mathbb{R}$ such that the restriction of A_1 to $\overline{\mathrm{lin}}\{T(t)x : t \geq 0\}$ is isomorphic to the multiplication operator $M_{is}f(s) := isf(s)$ on $L^2(\mathbb{R}, \mu_x)$. By Example 1.6, we see that

$$\lim_{t \to \infty} T(t)x = 0 \text{ weakly} \iff \mu_x \text{ is Rajchman.}$$

Note further that by Theorem 5.1 $T(\cdot)$ is almost weakly stable if and only if μ_x is continuous for every x. So we have the following.

Proposition 1.8. *A unitary C_0-group $U(\cdot)$ on a Hilbert space H with orthonormal basis S is weakly stable if and only if μ_x is Rajchman for every $x \in S$. Moreover, $U(\cdot)$ is almost weakly stable if and only if μ_x is continuous for every $x \in S$.*

1.2 Category result, discrete case

Inspired by the connection of stability and the Rajchman property, one can ask whether results analogous to the category results for operators and C_0-semigroups (see Sections II.5 and III.6) hold for measures, i.e., whether a "typical" Radon measure is continuous and not Rajchman for some appropriate topology. In fact, a much stronger fact holds, see also Nadkarni [193, Chapter 7].

Denote by $\mathcal{M}$ the set of all Radon probability measures on Γ. On this set we consider the weak* topology, i.e., $\lim_{n\to\infty} \mu_n = \mu$ if and only if $\lim_{n\to\infty} \int_\Gamma f\,d\mu_n = \int_\Gamma f\,d\mu$ for every $f \in C(\Gamma)$ or, equivalently, $\lim_{n\to\infty} \mu_n(A) = \mu(A)$ for every open set $A \subset \Gamma$. Since $\mathcal{M}$ is a closed subset of the closed unit ball of the dual space $(C(\Gamma))'$ endowed with the weak* topology, it is a complete metric (and also compact) space and hence a Baire space.

In the following we identify Γ with $[0,1)$ and a measure μ on Γ with its canonical image on $[0,1]$ again denoted by μ with convention $\mu(\{0\}) = \mu(\{1\})$. Further we write $\mathcal{M}_{\mathrm{rat}} := \mathrm{conv}\{\delta_\lambda : \lambda \in \mathbb{Q} \cap [0,1]\}$.

The following well-known approximation property will play a role later.

Proposition 1.9. *The set $\mathcal{M}_{\mathrm{rat}}$ is dense in $\mathcal{M}$.*

Proof. Take $\mu \in \mathcal{M}$, $\varepsilon > 0$, $n \in \mathbb{N}$, and arbitrary open sets $A_1,\ldots,A_n \subset [0,1]$. We have to find $\mu_r \in \mathcal{M}_{\mathrm{rat}}$ such that $|\mu(A_j) - \mu_r(A_j)| < \varepsilon$ holds for every $j = 1,\ldots,n$. It suffices to prove the statement if all A_j are disjoint open intervals of $[0,1]$. We now define for some $a_j \in A_j \cap \mathbb{Q}$

$$\mu_r := \mu(A_1)\delta_{a_1} + \ldots + \mu(A_n)\delta_{a_n} + (1 - \mu(A_1) - \ldots - \mu(A_n))\delta_0$$

Then $\mu_r \in \mathcal{M}_{\mathrm{rat}}$ and $\mu(A_j) = \mu_r(A_j)$ for every $j = 1,\ldots,n$. $\qquad\square$

We will also need an analogous statement for the set of all continuous measures denoted by $\mathcal{M}_{\mathrm{cont}}$.

Proposition 1.10. *The set $\mathcal{M}_{\mathrm{cont}}$ is dense in $\mathcal{M}$.*

Proof. Let $\mu \in \mathcal{M}$ be arbitrary, $\varepsilon > 0$ and $f_1,\ldots,f_n \in C_{\mathrm{per}}([0,1])$. We have to find a continuous probability measure ν with $|\langle f_j, (\mu - \nu)\rangle| < \varepsilon$ for every $j = 1,\ldots,n$. Denote by μ_{cont} the continuous and by μ_{discr} the discrete component of μ, assuming $\mu_{\mathrm{discr}} \neq 0$. Denote the set of atoms of μ_{discr} by $\{s_j\}$ which is at most countable. Moreover, by $\sum_j \mu(\{s_j\}) \leq 1$ we can assume without loss of generality that this set is finite. (Otherwise change the measure of the atoms with big indices appropriately).

Fix s_j. Since $f_1,\ldots,f_n$ are continuous, there exists $a > 0$ such that

$$|f_l(s_j) - \frac{1}{a}\int_{s_j}^{s_j+a} f_l\,d\lambda| < \varepsilon \quad \forall l = 1,\ldots,n,$$

where λ denotes Lebesgue's measure. (If $s_j = 1$, we take $\int_{1-a}^1 f_l\,d\mu$ instead of $\int_{s_j}^{s_j+a} f_l\,d\lambda$.) Moreover, we may assume that the intervals $[s_j, s_j + a]$ (respectively, $[1-a,1]$) are disjoint.

Define now $\nu_j := \frac{\mu(\{s_j\})}{a}\mathbf{1}_{[s_j,s_j+a]}\lambda$ and $\nu := \mu_{\mathrm{cont}} + \sum_j \nu_j$. Then ν is continuous and $\nu([0,1]) = \mu_{\mathrm{cont}}([0,1]) + \sum_j \mu(\{s_j\}) = \mu([0,1]) = 1$. Moreover, for

$l = 1, \ldots, n$ we have

$$|\langle f_l, (\mu - \nu)\rangle| = |\langle f_l, (\mu_{\text{discr}} - \sum_j \nu_j)\rangle| \leq \sum_j \left| f_l(s_j) - \frac{1}{a} \int_{s_j}^{s_j + a} f_l \, d\lambda \right| \mu(\{s_j\})$$

$$< \varepsilon \sum_j \mu(\{s_j\}) \leq \varepsilon,$$

and the proposition is proved. $\qquad\qquad\qquad\qquad\qquad\qquad\qquad\qquad\square$

We now introduce a class of measures having a much stronger property than being non-Rajchman.

Definition 1.11. A probability measure μ on the unit circle Γ is called *rigid* if its Fourier coefficients satisfy $\lim_{j \to \infty} \hat{\mu}_{n_j} = 1$ for some subsequence $\{n_j\}_{j=1}^\infty \subset \mathbb{N}$. Analogously, μ is called λ-*rigid* for some $\lambda \in \Gamma$ if there exists an increasing sequence $\{n_j\}_{j=1}^\infty \subset \mathbb{N}$ such that $\lim_{j \to \infty} \hat{\mu}_{n_j} = \lambda$. We finally call a measure Γ-*rigid* if it is λ-rigid for every $\lambda \in \Gamma$.

Note that, μ being a probability measure, one always has $|\hat{\mu}_n| \leq 1$ for all $n \in \mathbb{Z}$, so λ-rigidity is a kind of extremal property.

Remark 1.12. 1) For any $\lambda \in \Gamma$, λ-rigidity implies rigidity. This follows from $1 \in \{\lambda^n : n \in \mathbb{N}\}$ and the fact that the limit sets are closed.

2) If λ is irrational, i.e., satisfying $\lambda \notin e^{i\pi\mathbb{Q}}$, then every λ-rigid measure is automatically Γ-rigid, since $\{\lambda^n : n \in \mathbb{N}\}$ is dense in Γ.

3) As a converse to 1), there is an elegant argument how to produce λ-rigid (and hence Γ-rigid if λ is irrational) measures from a given rigid one via a rescaling argument, see Nadkarni [193, p. 50].

The easiest example of rigid measures are point measures. Indeed, a point measure δ_λ is always rigid, and it is Γ-rigid if and only if λ is irrational. Examples of singular rigid measures will be briefly discussed in Subsection 2.2.

We are now ready to prove the following category result due to Choksi, Nadkarni [51], see also Nadkarni [193, Chapter 7].

Theorem 1.13. *The set of all continuous rigid measures is residual in* $\mathcal{M}$. *In particular, the set* $\mathcal{M}_R$ *of all Rajchman measures is of first category, while the set* $\mathcal{M}_{cont}$ *of all continuous measures is residual.*

Proof. We start by proving that the set $\mathcal{M}_{\text{rig}}$ of all rigid measures is residual. Consider the sets

$$M_{k,n} = \{\mu : |\hat{\mu}_n - 1| < \frac{1}{k}\},$$

where $\hat{\mu}_n$ is the n-th Fourier coefficient of μ. These sets are open for the weak* topology by the definition of Fourier coefficients. Define furthermore the sets

$$N_{k,l} := \cup_{n \geq l} M_{k,n} = \{\mu : |\hat{\mu}_n - 1| < \frac{1}{k} \text{ for some } n \geq l\}$$

which are open as well. So we have that

$$\mathcal{M}_{\mathrm{rig}} = \cap_{k=1}^{\infty} N_{k,l}$$

is a G_δ set. Moreover, since $\mathcal{M}_{\mathrm{rat}} \subset \mathcal{M}_{\mathrm{rig}}$ and $\mathcal{M}_{\mathrm{rat}}$ is dense in $\mathcal{M}$ by Proposition 1.9, $\mathcal{M}_{\mathrm{rig}}$ is residual as a dense countable intersection of open sets.

We now prove that $\mathcal{M}_{\mathrm{cont}}$ is residual. Consider the countable set $Y := \mathrm{lin}_{\mathbb{Q}}\{e_n : n \in \mathbb{Z}\}$ of all linear combinations of $\{e_n\}$ with rational coefficients, where $e_n(s) := e^{2\pi i n s}$. By the Weierstrass approximation theorem, Y is dense in $C_{\mathrm{per}}([0,1])$. We numerate the elements of Y by $\{f_j\}_{j=1}^{\infty}$.

Define the open sets

$$W_{jkn} := \{\mu \in \mathcal{M} : |\langle e_n, f_j\mu\rangle| < \frac{1}{k}\}$$

(here and later $f\mu$ stands for the measure $f\, d\mu$) and

$$W_{jk} := \bigcup_{n=1}^{\infty} W_{jkn} = \{\mu \in \mathcal{M} : |\langle e_n, f_j\mu\rangle| < \frac{1}{k} \text{ for some } n \in \mathbb{N}\}.$$

We show

$$\mathcal{M}_{\mathrm{cont}} = \bigcap_{j,k=1}^{\infty} W_{jk}. \tag{IV.3}$$

By Proposition 1.2 we have the inclusion " $\subset$". To show the converse consider $\mu \notin \mathcal{M}_{\mathrm{cont}}$. Then there exists $s \in [0,1]$ with $\mu(\{s\}) =: d > 0$. For $f := 1_{\{s\}} \in L^1([0,1], \mu)$ we obtain

$$\langle e_n, f\mu\rangle = e^{2\pi i n s} d.$$

Since $\{f_j\}_{j=1}^{\infty}$ is dense in $L^1([0,1], \mu)$ (recall the convention $\mu(\{0\}) = \mu(\{1\})$), there is some $j \in \mathbb{N}$ such that $\int_0^1 |f_j - f|\, d\mu \leq d/2$. This implies

$$|\langle e_n, f_j\mu\rangle| \geq |\langle e_n, f\mu\rangle| - |\langle e_n, (f - f_j)\mu\rangle| \geq d - \frac{d}{2} = \frac{d}{2}$$

for all $n \in \mathbb{Z}$. So for an arbitrary fixed $k > 2/d$ we obtain that $\mu \notin W_{jk}$, therefore $\mu \notin \cap_{j,k=1}^{\infty} W_k$ and (IV.3) is proved. Thus W_{cont} is a dense countable intersection of open sets by the above and Proposition 1.10. This proves the assertion. $\square$

Analogously to the first part of the above proof and using a delicate rescaling argument mentioned in Remark 1.12, one can prove that for $\lambda \in \Gamma$, all λ-rigid measures are residual in $\mathcal{M}$ as well, see Nadkarni [193, 7.17]. Using this and Remark 1.12 2), we obtain the following result.

Theorem 1.14. *The set of measures μ satisfying*

(a) *μ is continuous, i.e., $\lim_{j\to\infty} \hat{\mu}_{n_j} = 0$ for some $\{n_j\} \subset \mathbb{Z}$ with density 1,*

(b) Γ *is contained in the limit set of the Fourier coefficients of* μ

is residual in $\mathcal{M}$ *for the weak* topology.*

Remark 1.15. One can add several other interesting properties to the above list of "typical" properties for measures. For example, the measures which are orthogonal to a given one are residual, which implies in particular that singular measures are residual. For a detailed discussion of this subject we refer to Nadkarni [193, Chapter 7].

1.3 Category result, continuous case

Analogous phenomena hold for probability measures on the real line. By $\mathcal{M}^{\mathbb{R}}$ we denote the set of all probability measures μ on $\mathbb{R}$ endowed with the weak* topology, i.e., $\mu_n \to \mu$ if and only if $\langle f, \mu_n \rangle \to \langle f, \mu \rangle$ for every $f \in C_b(\mathbb{R})$. Since $\mathcal{M}^{\mathbb{R}}$ is a closed subset of the unit ball in $(C_b(\mathbb{R}))'$ for the weak* topology, it is a complete metric (and compact) space.

By just the same arguments as in the discrete case, we obtain the following.

Theorem 1.16. *The set of measures* μ *satisfying*

(a) μ *is continuous,*

(b) μ *is* Γ*-rigid, i.e.,* Γ *is contained in the limit set of the Fourier transform of* μ

is residual in $\mathcal{M}^{\mathbb{R}}$ *for the weak* topology. In particular, the set of all Rajchman measures on* $\mathbb{R}$ *is of first category, while the set of all continuous measures is residual.*

In Example 3.21 b) below we briefly discuss abstract examples of such measures coming from weakly mixing Γ-rigid automorphisms.

2 Stability and mixing

We now turn our attention to the analogue of stability in ergodic theory.

Ergodic theory deals with a probability space (Ω, Σ, μ) and a *measure preserving transformation* (m.p.t.) $\varphi : \Omega \to \Omega$, i.e., a measurable map satisfying $\mu(\varphi^{-1}(B)) = \mu(B)$ for all $B \in \Sigma$. The continuous analogue is a *measure preserving semiflow* (m.p. semiflow) $(\varphi_t)_{t \geq 0}$, where each φ_t is a m.p.t., the function $(\omega, t) \mapsto \varphi_t(\omega)$ is measurable and $\varphi_0 = Id$, $\varphi_{t+s} = \varphi_t \varphi_s$ holds for all $t, s \geq 0$. We refer to Cornfeld, Fomin, Sinai [53], Krengel [154], Petersen [212], Halmos [118], or Tao [240, Chapter II] for basic facts and further information.

To translate the above concepts into operator theoretic language, fix $1 \leq p < \infty$ and define $X := L^p(\Omega, \Sigma, \mu)$ and $T \in \mathcal{L}(X)$ by $Tf := f \circ \varphi$, $f \in L^p(\Omega, \mu)$. Then T is a *linear* operator on X which is isometric since, φ being measure preserving,

$$\|Tf\|^p = \int_\Omega |f(\varphi(\omega))|^p \, d\mu = \int_\Omega |f(\omega)|^p \, d\mu = \|f\|^p.$$

Thus, for φ invertible and $p = 2$, we obtain a unitary operator T. Note that the operator T has relatively weakly compact orbits for $p > 1$ by the theorem of Banach-Alaoglu and for $p = 1$ by Example I.1.7 (b) with $u = 1$.

Analogously, for a m.p. semiflow (φ_t) define $(T(t))_{t\geq 0}$ by $T(t)f := f \circ \varphi_t$, $f \in L^p(\Omega, \mu)$. Then $(T(t))_{t\geq 0}$ is a semigroup of *linear* isometries which is strongly continuous by Krengel [154], §1.6, Thm. 6.13. Thus, if one/each φ_t is invertible (i.e., we start with a flow) and $p = 2$, then $T(\cdot)$ becomes a unitary C_0-group. Moreover, $T(\cdot)$ is relatively weakly compact (for $p = 1$ use again Example I.1.7 (b) with $u = 1$).

We call T the *induced operator* and $T(\cdot)$ the *induced C_0-semigroup*. This operator (and each operator of the C_0-semigroup) has remarkable additional properties: it is multiplicative on the Banach algebra $L^\infty(\Omega, \mu)$ and positive on each of the Banach lattices $L^p(\Omega, \mu)$, i.e., it maps positive functions into positive functions. This implies, as an easy consequence of the Perron–Frobenius theory for such operators (see Schaefer [227, Section V.4]), that the boundary spectrum $\sigma(T) \cap \Gamma$ as well as the boundary point spectrum $P_\sigma(T) \cap \Gamma$ are *cyclic* subsets of Γ, i.e., $\lambda \in \sigma(T) \cap \Gamma$ implies $\lambda^n \in \sigma(T) \cap \Gamma$ for all $n \in \mathbb{Z}$, and analogously for $P_\sigma(T)$.

2.1   Ergodicity

We begin with ergodicity, the basic notion in ergodic theory.

Discrete case

For a m.p.t. φ on (Ω, Σ, μ), a set $B \in \Sigma$ is called φ-*invariant* if $\mu(\varphi^{-1}(B) \triangle B) = 0$, and φ is *ergodic* if for every φ-invariant set B either $\mu(B) = 0$ or $\mu(B) = 1$ holds. In other words, ergodic transformations leave no non-trivial set invariant.

Translating this property into a property of the induced operator leads to the following.

Theorem 2.1. *A m.p.t. φ is ergodic if and only if the induced operator satisfies* $\mathrm{Fix}\,T = \langle 1 \rangle$. *In this case every eigenvalue of T is simple.*

Proof. Note first that the constant functions are fixed under T. Assume $\mathrm{Fix}\,T = \langle 1 \rangle$. If $B \in \Sigma$ is φ-invariant, we obtain $T1_B = 1_B$, hence $1_B = 1$ or $1_B = 0$ by assumption, and thus φ is ergodic. Assume now that φ is ergodic and take $f \in \mathrm{Fix}\,T$. Since each set $B_r := \{\omega \in \Omega : \mathrm{Re}\, f(\omega) > r\}$, $r \in \mathbb{R}$, is φ-invariant, it must be trivial, hence $\mathrm{Re}\, f$ is constant a.e.. Analogously, $\mathrm{Im}\, f$ is constant and thus $f \in \langle 1 \rangle$. The first part of the theorem is proved.

To show the second part take $\lambda \in P_\sigma(T)$ and $0 \neq f \in X$ with $Tf = f \circ \varphi = \lambda f$. Then $T\overline{f} = \overline{\lambda} \cdot \overline{f}$ and, since T is isometric, $\lambda \in \Gamma$ and therefore $T|f| = |f|$. Thus $|f|$ is a non-zero constant function by $\mathrm{Fix}\,T = \langle 1 \rangle$. Assume $|f| = 1$ and take another eigenfunction g corresponding to the eigenvalue λ with $|g| = 1$. Then $T(g\overline{f}) = g\overline{f}$ implies $g\overline{f} = 1$, so $f = g$ and λ is simple. $\square$

By the mean ergodic theorem applied to the induced operator T, ergodicity can be reformulated into an asymptotic property of the transformation φ and the operator T.

Corollary 2.2. *For a m.p.t. φ on (Ω, Σ, μ), $1 \leq p < \infty$, and the induced operator T on $X = L^p(\Omega, \mu)$, the following assertions are equivalent.*

(i) φ *is ergodic.*

(ii) $\lim\limits_{N \to \infty} \dfrac{1}{N+1} \sum\limits_{n=0}^{N} \mu(\varphi^{-n}(B) \cap C) = \mu(B)\mu(C)$ *for every* $B, C \in \Sigma$.

(iii) $\lim\limits_{N \to \infty} \dfrac{1}{N+1} \sum\limits_{n=0}^{N} T^n f = \int_{\Omega} f \, d\mu \cdot \mathbf{1}$ *for every* $f \in X$.

Proof. (i)$\Rightarrow$(iii) As mentioned above, T has relatively weakly compact orbits and is therefore mean ergodic by Theorem I.2.9. If P is the mean ergodic projection of T and $f \in X$, we obtain by ergodicity of φ that $Pf = C\mathbf{1}$ for some constant C by Proposition I.2.8. To determine this constant observe that

$$\lim_{N \to \infty} \frac{1}{N+1} \sum_{n=0}^{N} \int_{\Omega} T^n f \, d\mu = \int_{\Omega} f \, d\mu,$$

where we used that φ preserves the measure μ. This implies $C = \int_{\Omega} f d\mu$, and the claim follows.

(iii)$\Rightarrow$(ii) For $B, C \in \Sigma$ and $f := \mathbf{1}_B$, we have by (iii),

$$\frac{1}{N+1} \sum_{n=0}^{N} \mu(\varphi^{-n}(B) \cap C) = \frac{1}{N+1} \sum_{n=0}^{N} \int_{C} \mathbf{1}_{\varphi^{-n}(B)} \, d\mu = \frac{1}{N+1} \sum_{n=0}^{N} \int_{C} T^n f \, d\mu$$

$$\to \int_{\Omega} \mathbf{1}_B \, d\mu \int_{C} \mathbf{1} \, d\mu = \mu(B)\mu(C)$$

as $N \to \infty$.

To show (ii)$\Rightarrow$(i), take B with $\mu(\varphi^{-1}(B) \triangle B) = 0$. Property (ii) implies

$$\mu(B)^2 = \lim_{N \to \infty} \frac{1}{N+1} \sum_{n=0}^{N} \mu(\varphi^{-n}(B) \cap B) = \mu(B).$$

Hence either $\mu(B) = 0$ or $\mu(B) = 1$ showing ergodicity of φ. $\qquad \square$

It follows from the proof that it suffices to take $C = B$ in (ii).

Remark 2.3. One can interpret (iii) as "time mean equals space mean" which is a version of the famous *ergodic hypothesis* by Boltzmann from around 1880 (see, e.g., Halmos [118, Introduction] and [68, Chapter 1]).

Continuous case

Ergodicity for semiflows is defined analogously, and we state the relevant properties without proof, see, e.g., Cornfeld, Fomin, Sinai [53, §1.2, 1.4].

A measure preserving semiflow $(\varphi_t)_{t\geq0}$ is called *ergodic* if the only invariant sets under all φ_t are trivial, i.e., if $\mu(\varphi_t^{-1}(B)\triangle B) = 0$ for every $t \geq 0$ implies either $\mu(B) = 0$ or $\mu(B) = 1$. Again, ergodicity can be characterised in terms of the induced C_0-semigroup $T(\cdot)$.

Theorem 2.4. *Let φ be a measure preserving semiflow and $T(\cdot)$ the induced C_0-semigroup with generator A on $L^p(\Omega, \mu)$, $1 \leq p < \infty$. Then φ is ergodic if and only if* $\operatorname{Fix} T(\cdot) := \bigcap_{t\geq0} \operatorname{Fix} T(t) = \ker A = \langle 1 \rangle$. *In this case every eigenvalue of A is simple.*

Analogously to the discrete case, mean ergodicity of $T(\cdot)$ allows the following characterisation of ergodicity of $(\varphi_t)_{t\geq0}$.

Corollary 2.5. *For a measure preserving semiflow $(\varphi_t)_{t\geq0}$ and the induced semigroup $T(\cdot)$ on $L^p(\Omega, \mu)$, $1 \leq p < \infty$, the following assertions are equivalent.*

(i) $(\varphi_t)_{t\geq0}$ *is ergodic.*

(ii) $\displaystyle\lim_{t\to\infty} \frac{1}{t} \int_0^t \mu(\varphi_s^{-1}(B) \cap C)ds = \mu(B)\mu(C)$ *for every $B, C \in \Sigma$.*

(iii) $\displaystyle\lim_{t\to\infty} \frac{1}{t} \int_0^t T(s)f ds = \int_\Omega f d\mu \cdot 1$ *for every $f \in L^p(\Omega, \mu)$.*

Again one can take $C = B$ in (ii).

We see that, in the discrete and in the continuous case, ergodicity is equivalent to an asymptotic property of φ expressed by assertions (ii) and (iii) in Corollaries 2.2 and 2.5. Property (ii) means that $\varphi_t^{-1}(A)$ becomes "equidistributed" in the Cesàro sense. Stronger asymptotic properties are discussed below.

2.2 Strong and weak mixing

The so-called "mixing properties" in ergodic theory are closely related to weak and almost weak stability as studied in Sections II.4 and III.5. We explain this relation now.

Discrete case

A measure preserving transformation φ on a probability space (Ω, Σ, μ) is called *strongly mixing* if

$$\lim_{n\to\infty} \mu(\varphi^{-n}(B) \cap C) = \mu(B)\mu(C)$$

for any two measurable sets $B, C \in \Sigma$. The transformation φ is called *weakly mixing* if

$$\lim_{N \to \infty} \frac{1}{N+1} \sum_{n=0}^{N} |\mu(\varphi^{-n}(B) \cap C) - \mu(B)\mu(C)| = 0$$

holds for all B, C.

These concepts play an important role in ergodic theory, and we refer to Cornfeld, Fomin, Sinai [53], Krengel [154], Petersen [212], Walters [254] or Halmos [118] for further information.

Strong mixing implies weak mixing and weak mixing implies ergodicity by Corollary 2.2, but the converse implications do not hold in general. A famous result of Halmos [116] and Rohlin [221] states that a "typical" m.p.t. (for the strong operator topology for the induced operators) is weakly but not strongly mixing. However, explicit examples of weakly but not strongly mixing transformations are not easy to construct, see e.g. Lind [171] for a concrete example and Petersen [212, p. 209] for a method of constructing such transformations. An important class of weakly but not strongly mixing transformations are so-called rigid transformations discussed briefly in Example 3.12 b) below.

We first translate the above concepts into the operator-theoretic language (see, e.g., Halmos [118, pp. 37–38]).

Proposition 2.6. *Let φ be a m.p.t. on a probability measure space (Ω, Σ, μ), $1 \le p < \infty$, and let T be the induced linear operator on $X := L^p(\Omega, \mu)$ defined by $Tf := f \circ \varphi$. Denote further by P the projection given by $Pf := \int_\Omega f \, d\mu \cdot \mathbf{1}$, $f \in X$. Then the following assertions hold.*

(a) *φ is strongly mixing if and only if*

$$\lim_{n \to \infty} \langle T^n f, g \rangle = \langle Pf, g \rangle \quad \text{for all } f \in X, \ g \in X'.$$

(b) *φ is weakly mixing if and only if*

$$\lim_{N \to \infty} \frac{1}{N+1} \sum_{n=0}^{N} |\langle T^n f, g \rangle - \langle Pf, g \rangle| = 0 \quad \text{for all } f \in X, \ g \in X'.$$

Proof. (a) By definition, φ is strongly mixing if and only if

$$\lim_{n \to \infty} \int_\Omega T^n \mathbf{1}_B \cdot \mathbf{1}_C \, d\mu = \int_\Omega \mathbf{1}_B \, d\mu \cdot \int_\Omega \mathbf{1}_C \, d\mu$$

for every $B, C \in \Sigma$, which, by a standard density argument in the Banach space X, is equivalent to

$$\lim_{n \to \infty} \int_\Omega T^n f \cdot g \, d\mu = \int_\Omega f \, d\mu \cdot \int_\Omega g \, d\mu$$

for every $f \in X$ and $g \in X'$.

(b) Analogously to the above, φ is weakly mixing if and only if

$$\lim_{N \to \infty} \frac{1}{N+1} \sum_{n=0}^{N} |\langle T^n f, g \rangle - \langle Pf, g \rangle| = \lim_{N \to \infty} \frac{1}{N+1} \sum_{n=0}^{N} |\langle (T^n - P)f, g \rangle| = 0$$

holds for every $f := \mathbf{1}_B$ and $g := \mathbf{1}_C$ with $B, C \in \Sigma$. By the triangle inequality, this is equivalent to the same property for every f, g in the linear hull of characteristic functions and, by the standard density argument, for every $f \in X$ and $g \in X'$. $\quad\square$

In order to relate mixing to stability properties, i.e., convergence to 0, we only need to eliminate the constant functions which are always fixed under T. To this end consider the decomposition $X = \langle \mathbf{1} \rangle \oplus X_0$, where

$$X_0 := \left\{ f \in X : \int_\Omega f \, d\mu = 0 \right\}$$

is closed and T-invariant. Note that the restriction $T_0 := T|_{X_0}$ has relatively weakly compact orbits as well.

The following connects mixing properties of φ to stability properties of T_0 and follows directly from the above proposition.

Corollary 2.7. *For a m.p.t. φ on a probability space (Ω, Σ, μ) and the corresponding operator T_0 on $X_0 \subset L^p(\Omega, \mu)$, $1 \le p < \infty$, the following assertions hold.*

(a) *φ is strongly mixing if and only if T_0 is weakly stable.*

(b) *φ is weakly mixing if and only if T_0 is almost weakly stable.*

Remark 2.8. Every weakly but not strongly mixing transformation induces an almost weakly but not weakly stable operator. However, as mentioned above, an explicit example of such a transformation and hence of such an operator is not easy to construct, see e.g. Petersen [212, Section 4.5], Example 3.12 b) and Example 3.17 below.

Continuous case

We now give the continuous analogues of the above concepts and present some results connecting mixing flows to stable C_0-semigroups.

A measure preserving semiflow $(\varphi_t)_{t \ge 0}$ on a probability space (Ω, Σ, μ) is called *strongly mixing* (or just *mixing*) if

$$\lim_{t \to \infty} \mu(\varphi_t^{-1}(B) \cap C) = \mu(B)\mu(C)$$

holds for every $B, C \in \Sigma$. The semiflow $(\varphi_t)_{t \ge 0}$ is called *weakly mixing* if for all $B, C \in \Sigma$ we have

$$\lim_{t \to \infty} \frac{1}{t} \int_0^t |\mu(\varphi_s^{-1}(B) \cap C) - \mu(B)\mu(C)| \, ds = 0.$$

Strong mixing implies weak mixing and weak mixing implies ergodicity, but the converse implications do not hold in general. For an example of a weakly mixing flow which is not mixing see, e.g., Cornfeld, Fomin, Sinai [53, Chapter 14], Katok [141], Katok, Stepin [144], Avila, Forni [11], Ulcigrai [244], Scheglov [228], Kulaga [159] and also Example 3.21 b) below. Note that there seems to be no continuous analogue of Halmos' and Rohlin's result on "typicality" of weakly but not strongly mixing flows.

The following continuous analogue of Proposition 2.6 holds.

Proposition 2.9. *Let* $(\varphi_t)_{t\geq 0}$ *be a measurable measure preserving semiflow on a probability measure space* (Ω, Σ, μ), $1 \leq p < \infty$, *and let* $T(\cdot)$ *be the induced* C_0-*semigroup on* $X = L^p(\Omega, \mu)$. *Let further* P *be the projection given by* $Pf := \int_\Omega f \, d\mu \cdot \mathbf{1}$, $f \in X$. *Then the following assertions hold.*

(a) $(\varphi_t)_{t\geq 0}$ *is strongly mixing if and only if*

$$\lim_{t\to\infty} \langle T(t)f, g\rangle = \langle Pf, g\rangle \quad \text{for all } f \in X, \ g \in X'.$$

(b) $(\varphi_t)_{t\geq 0}$ *is weakly mixing if and only if*

$$\lim_{t\to\infty} \frac{1}{t} \int_0^t |\langle T(s)f, g\rangle - \langle Pf, g\rangle| \, ds = 0 \quad \text{for all } f \in X, \ g \in X'.$$

As before, the constant functions are fixed under the induced C_0-semigroup. So we again consider the decomposition $X = \langle \mathbf{1}\rangle \oplus X_0$ for the closed and $(T(t))_{t\geq 0}$-invariant subspace

$$X_0 := \left\{ f \in X : \int_\Omega f \, d\mu = 0 \right\}.$$

We denote the restriction of $(T(t))_{t\geq 0}$ to X_0 by $(T_0(t))_{t\geq 0}$ and its generator by A_0. The semigroup $(T_0(t))_{t\geq 0}$ remains relatively weakly compact.

This leads to the following characterisation of mixing of $(\varphi_t)_{t\geq 0}$ in terms of stability of the semigroup $(T_0(t))_{t\geq 0}$.

Corollary 2.10. *For a measure preserving semiflow* $(\varphi_t)_{t\geq 0}$ *on a probability space* (Ω, Σ, μ), $1 \leq p < \infty$, *and the corresponding semigroup* $T_0(\cdot)$ *on* $X_0 \subset L^p(\Omega, \mu)$, *the following assertions hold.*

(a) $(\varphi_t)_{t\geq 0}$ *is strongly mixing if and only if* $T_0(\cdot)$ *is weakly stable.*

(b) $(\varphi_t)_{t\geq 0}$ *is weakly mixing if and only if* $T_0(\cdot)$ *is almost weakly stable.*

So every weakly but not strongly mixing flow yields an almost weakly but not weakly stable C_0-semigroup.

3   Rigidity phenomena

Inspired by results in ergodic and measure theory, we now describe the "typical" (in the Baire category sense) asymptotic behaviour of unitary, isometric and contractive operators on separable Hilbert spaces using the notion of rigidity. We will see that the "typical" behaviour is quite counterintuitive and can be viewed as "random": The strong limit set of the powers of a "typical" contraction contains the unit circle times the identity operator, while most of the powers converge weakly to zero. We also give the continuous analogue for unitary and isometric C_0-(semi)groups. This in particular generalises category results from Sections II.4 and III.5.

3.1   Preliminaries

We first define rigid and Γ-rigid operators and C_0-semigroups and make some elementary observations.

Definition 3.1. A bounded operator T on a Banach space X is called *rigid* if

$$\text{strong-}\lim_{j\to\infty} T^{n_j} = I \quad \text{for some subsequence } \{n_j\}_{j=1}^{\infty} \subset \mathbb{N}.$$

Note that in the above definition one can remove the assumption $\lim_{j\to\infty} n_j = \infty$. (Indeed, if n_j does not converge to ∞, then $T^{n_0} = I$ for some n_0 implying $T^{nn_0} = I$ for every $n \in \mathbb{N}$.)

Remark 3.2. Rigid operators have no non-trivial weakly stable orbit. In particular, by the Foiaş–Sz.-Nagy decomposition (see Theorem II.3.9), rigid contractions on Hilbert spaces are necessarily unitary.

As trivial examples of rigid operators take $T := \lambda I$ for $|\lambda| = 1$. Moreover, arbitrary (countable) combinations of such operators are rigid as well, as the following proposition shows.

Proposition 3.3. *Let X be a separable Banach space and let $T \in \mathcal{L}(X)$ be power bounded with discrete spectrum, i.e., satisfying*

$$H = \overline{\lim}\{x \in X : Tx = \lambda x \text{ for some } \lambda \in \Gamma\}.$$

Then T is rigid.

Proof. Since X is separable, the strong operator topology is metrisable on bounded sets of $\mathcal{L}(H)$ (take for example the metric $d(T, S) := \sum_{j=0}^{\infty} \|Tz_j - Sz_j\|/(2^j\|z_j\|)$ for a dense sequence $\{z_j\}_{j=1}^{\infty} \subset X \setminus \{0\}$). So it suffices to show that I belongs to the strong closure of $\{T^n\}_{n\in\mathbb{N}}$. For $\varepsilon > 0$, $m \in \mathbb{N}$ and $x_1, \ldots, x_m \in X$, we have to find $n \in \mathbb{N}$ such that $\|T^n x_j - x_j\| < \varepsilon$ for every $j = 1, \ldots, m$.

Assume first that each x_j is an eigenvector with $\|x_j\| = 1$ corresponding to some unimodular eigenvalue λ_j, and hence $\|T^n x_j - x_j\| = |\lambda_j^n - 1|\|x\|$. Consider the compact group Γ^m and the rotation $\varphi : \Gamma^m \to \Gamma^m$ given by $\varphi(z) := az$ for $a := (\lambda_1, \ldots, \lambda_m)$. By a classical recurrence theorem, see e.g. Furstenberg [92, Theorem 1.2], there exists n such that $|\varphi^n(1) - 1| < \varepsilon$, i.e.,

$$|\lambda_j^n - 1| < \varepsilon \quad \text{for every } j = 1, \ldots, m,$$

implying $\|T^n x_j - x_j\| < \varepsilon$ for every $j = 1, \ldots, m$.

Assume now $0 \neq x_j \in \lin\{x \in X : Tx = \lambda x \text{ for some } \lambda \in \Gamma\}$ for all j. Then we have $x_j = \sum_{k=1}^K c_{jk} y_k$ for $K \in \mathbb{N}$, eigenvectors $y_k \in X$ with $\|y_k\| = 1$, and $c_{jk} \in \mathbb{C}$, $k = 1, \ldots, K$, $j = 1, \ldots, m$. Take $\varepsilon_2 := \frac{\varepsilon}{K \max_{j,k} |c_{jk}|}$. By the above, there exists $n \in \mathbb{N}$ such that $\|T^n y_k - y_k\| < \varepsilon_2$ for every $k = 1, \ldots, K$ and therefore

$$\|T x_j - x_j\| \leq \sum_{k=1}^K |c_{jk}| \|T y_k - y_k\| < \varepsilon$$

for every $j = 1, \ldots, m$. The standard density argument covers the case of arbitrary $x_j \in X$. $\qquad\square$

Analogously, one defines λ-rigid operators by replacing I by λI in the above definition.

Definition 3.4. Let X be a Banach space, $T \in \mathcal{L}(X)$ and $\lambda \in \Gamma$. We call T λ-*rigid* if there exists a subsequence $\{n_j\}_{j=1}^\infty \subset \mathbb{N}$ such that

$$\text{strong-} \lim_{j \to \infty} T^{n_j} = \lambda I.$$

Again one can remove the assumption $\lim n_j = \infty$. Finally, T is Γ-*rigid* if T is λ-rigid for every $\lambda \in \Gamma$.

Remark 3.5. Since every λ-rigid operator is λ^n-rigid for every $n \in \mathbb{N}$, we see that λ-rigidity implies rigidity. Moreover, λ-rigidity is equivalent to Γ-rigidity whenever λ is irrational, i.e., $\lambda \notin e^{2\pi i \mathbb{Q}}$. (We used the fact that for irrational λ the set $\{\lambda^n\}_{n=1}^\infty$ is dense in Γ and that limit sets are always closed.)

The simplest examples are again operators of the form λI, $|\lambda| = 1$. Indeed, $T = \lambda I$ is λ-rigid, and it is Γ-rigid if and only if λ is irrational.

Analogously, one can define rigidity for strongly continuous semigroups.

Definition 3.6. A C_0-semigroup $(T(t)_{t \geq 0})$ on a Banach space is called λ-*rigid* for $\lambda \in \Gamma$ if there exists a sequence $\{t_j\}_{j=1}^\infty \subset \mathbb{R}_+$ with $\lim_{j \to \infty} t_j = \infty$ such that

$$\text{strong-} \lim_{j \to \infty} T(t_j) = \lambda I.$$

Semigroups which are 1-rigid are called *rigid*, and if they are rigid for every $\lambda \in \Gamma$ they are called Γ-*rigid*.

As in the discrete case, λ-rigidity for some λ implies rigidity, and λ-rigidity for some irrational λ is equivalent to Γ-rigidity.

Remark 3.7. Again, rigid semigroups have no non-zero weakly stable orbit. This implies by the Foiaş–Sz.-Nagy decomposition, see Theorem III.4.7, that every rigid semigroup on a Hilbert space is automatically unitary.

The simplest examples of rigid C_0-semigroups are given by $T(t) = e^{iat}I$ for some $a \in \mathbb{R}$. In this case, $T(\cdot)$ is automatically Γ-rigid whenever $a \neq 0$. Moreover, one has the following continuous analogue of Proposition 3.3.

Proposition 3.8. *Let X be a separable Banach space and let $T(\cdot)$ be a bounded C_0-semigroup with discrete spectrum, i.e., satisfying*

$$H = \overline{\lin}\{x \in X : T(t)x = e^{ita}x \text{ for some } a \in \mathbb{R} \text{ and all } t \geq 0\}.$$

Then $T(\cdot)$ is rigid.

So rigidity becomes non-trivial for operators (C_0-semigroups) having no point spectrum on the unit circle. Recall that for operators with relatively compact orbits, the absence of point spectrum on Γ is equivalent to almost weak stability, i.e.,

$$\text{weak-}\lim_{j\to\infty} T^{n_j} = 0 \text{ for some subsequence } \{n_j\} \text{ with density } 1,$$

and analogously for C_0-semigroups, see Theorems II.4.1 and III.5.1.

Restricting ourselves to Hilbert spaces, we will see that almost weakly stable and Γ-rigid operators and semigroups are the rule and not just an exception.

3.2   Discrete case: powers of operators

In this section we describe the "typical" (in the Baire category sense) asymptotic behaviour of unitary, isometric and contractive operators on separable Hilbert spaces.

As in Section II.5, we take a separable infinite-dimensional Hilbert space H and denote by $\mathcal{U}$ the set of all unitary operators on H endowed with the strong* operator topology. The set of all isometric operators on H with the strong operator topology is denoted by $\mathcal{I}$, and $\mathcal{C}$ will be the space of all contractions on H with the weak operator topology. All these spaces are complete metric and hence Baire.

We now prove the first but basic step to describe the asymptotics of contractions.

Theorem 3.9. *Let H be a separable infinite-dimensional Hilbert space. The set*

$$M := \{T : \lim_{j\to\infty} T^{n_j} = I \text{ strongly for some } n_j \to \infty\}$$

is residual for the weak operator topology in the set $\mathcal{C}$ of all contractions on H. This set is also residual for the strong operator topology in the set $\mathcal{I}$ of all isometries and in the set $\mathcal{U}$ of all unitary operators for the strong* operator topology.

Proof. We begin with the isometric case.

Let $\{x_l\}_{l=1}^\infty$ be a dense subset of $H\backslash\{0\}$. Since one can remove the assumption $\lim_{j\to\infty} n_j = \infty$ from the definition of M, we have

$$M = \{T \in \mathcal{I}:\ \exists\{n_j\}_{j=1}^\infty \subset \mathbb{N} \text{ with } \lim_{j\to\infty} T^{n_j} x_l = x_l\ \forall l \in \mathbb{N}\}. \qquad \text{(IV.4)}$$

Consider the sets

$$M_k := \left\{ T \in \mathcal{I}:\ \sum_{l=1}^\infty \frac{1}{2^l \|x_l\|} \|T^n x_l - x_l\| < \frac{1}{k} \text{ for some } n \right\}$$

which are open in the strong operator topology. Therefore,

$$M = \bigcap_{k=1}^\infty M_k$$

implies that M is a G_δ-set. To show that M is residual it just remains to prove that M is dense. Since M contains all periodic unitary operators which are dense in $\mathcal{I}$ by Proposition II.5.7, the assertion follows.

While for unitary operators the above arguments work as well, we need to be more careful in the space $\mathcal{C}$ of all contractions.

We first show that

$$M = \{T \in \mathcal{C}:\ \exists\{n_j\} \subset \mathbb{N} \text{ with } \lim_{j\to\infty} \langle T^{n_j} x_l, x_l \rangle = \|x_l\|^2\ \forall l \in \mathbb{N}\}. \qquad \text{(IV.5)}$$

The inclusion "$\subset$" is clear. To prove the converse inclusion, assume that $\lim_{j\to\infty} \langle T^{n_j} x_l, x_l \rangle = \|x_l\|^2$ for each $l \in \mathbb{N}$. By the standard density argument we have $\lim_{j\to\infty} \langle T^{n_j} x, x \rangle = \|x\|^2$ for every $x \in H$, and strong convergence of T^{n_j} to I follows from

$$\|(T^{n_j} - I)x\|^2 = \|T^{n_j}x\|^2 - 2\mathrm{Re}\,\langle T^{n_j}x, x \rangle + \|x\|^2 \le 2(\|x\|^2 - \langle T^{n_j}x, x \rangle).$$

We now define

$$M_k := \left\{ T \in \mathcal{C}:\ \sum_{l=1}^\infty \frac{1}{2^l \|x_l\|^2} |\langle (T^n - I)x_l, x_l \rangle| < \frac{1}{k} \text{ for some } n \right\},$$

and observe again that $M = \bigcap_k M_k$.

It remains to show that the complement M_k^c of M_k is a nowhere dense set. Since the set of periodic unitary operators U_{per} on H is dense in the set of all contractions for the weak operator topology (see Proposition II.5.10), it suffices to show $U_{per} \cap \overline{M_k^c} = \emptyset$. Assume that this is not the case, i.e., that there exists a sequence $\{T_m\}_{m=1}^\infty \subset M_k^c$ converging weakly to a periodic unitary operator U.

Then by Lemma II.5.11 we have $\lim_{m \to \infty} T_m = U$ strongly, hence $\lim_{m \to \infty} T_m^n = U^n$ strongly for every $n \in \mathbb{N}$. However, $T_m \in M_k^c$ means that

$$\sum_{l=1}^{\infty} \frac{1}{2^l \|x_l\|^2} |\langle (T_m^n - I)x_l, x_l \rangle| \geq \frac{1}{k} \quad \text{for every } n, m \in \mathbb{N}.$$

Since T_m^n converges strongly and hence weakly to U^n for every n, and hence the lth summand on the left-hand side of the above inequality is dominated by $\frac{1}{2^{l-1}}$ which form a sequence in l^1, we obtain by letting $m \to \infty$ that

$$\sum_{l=1}^{\infty} \frac{1}{2^l \|x_l\|^2} |\langle (U^n - I)x_l, x_l \rangle| \geq \frac{1}{k} \quad \text{for every } n,$$

contradicting the periodicity of U. □

We now show that one can replace I in Theorem 3.9 by λI for any $\lambda \in \Gamma$. To this purpose we need the following modification of Proposition II.5.1.

Lemma 3.10. *Let H be a Hilbert space, $\lambda \in \Gamma$ and $N \in \mathbb{N}$. Then the set of all unitary operators U with $U^n = \lambda I$ for some $n \geq N$ is dense in the set of all unitary operators for the norm topology.*

Proof. Let U be a unitary operator, $\lambda = e^{i\alpha} \in \Gamma$, $N \in \mathbb{N}$ and $\varepsilon > 0$. By the spectral theorem U is unitarily equivalent to a multiplication operator $\tilde{U}$ on some $L^2(\Omega, \mu)$ with

$$(\tilde{U}f)(\omega) = \varphi(\omega)f(\omega), \quad \forall \omega \in \Omega,$$

for some measurable $\varphi : \Omega \to \Gamma := \{z \in \mathbb{C} : |z| = 1\}$.

We now approximate the operator $\tilde{U}$ as follows. Take $n \geq N$ such that $|1 - e^{\frac{2\pi i}{n}}| \leq \varepsilon$ and define for $\alpha_j := e^{i\left(\frac{\alpha}{n} + \frac{2\pi j}{n}\right)}$, $j = 0, \ldots, n$,

$$\psi(\omega) := \alpha_{j-1}, \quad \forall \omega \in \varphi^{-1}(\{z \in \Gamma : \arg(\alpha_{j-1}) \leq \arg(z) < \arg(\alpha_j)\}).$$

The multiplication operator $\tilde{P}$ corresponding to ψ satisfies $\tilde{P}^n = e^{i\alpha}$. Moreover,

$$\|\tilde{U} - \tilde{P}\| = \sup_{\omega \in \Omega} |\varphi(\omega) - \psi(\omega)| \leq \varepsilon$$

proving the assertion. □

We now describe the "typical" asymptotic behaviour of contractions (isometries, unitary operators) on separable Hilbert spaces. For an alternative proof in the unitary case based on the spectral theorem and an analogous result for measures on Γ see Nadkarni [193, Chapter 7].

Theorem 3.11. *Let H be a separable infinite-dimensional Hilbert space. Then the set of all operators T satisfying the properties*

(1) *there exists $\{n_j\}_{j=1}^{\infty} \subset \mathbb{N}$ with density 1 such that*

$$\lim_{j \to \infty} T^{n_j} = 0 \quad weakly,$$

(2) *for every $\lambda \in \Gamma$ there exists $\{n_j^{(\lambda)}\}_{j=1}^{\infty}$ with $\lim_{j \to \infty} n_j^{(\lambda)} = \infty$ such that*

$$\lim_{j \to \infty} T^{n_j^{(\lambda)}} = \lambda I \quad strongly$$

is residual for the weak operator topology in the set $\mathcal{C}$ of all contractions. This set is also residual for the strong operator topology in the set $\mathcal{I}$ of all isometries as well as for the strong* operator topology in the set $\mathcal{U}$ of all unitary operators.

Recall that every contraction satisfying (2) is unitary, cf. Remark 3.2.

Proof. By Theorems II.5.4, II.5.9, II.5.12, operators satisfying (1) are residual in $\mathcal{C}, \mathcal{I}$ and $\mathcal{U}$.

We now show that for a fixed $\lambda \in \Gamma$, the set M of all operators T satisfying $\lim_{j \to \infty} T^{n_j} = \lambda I$ strongly for some sequence $\{n_j\}_{j=1}^{\infty}$ is residual. We again prove this first for isometries and the strong operator topology.

Take a dense set $\{x_l\}_{l=1}^{\infty}$ of $H \setminus \{0\}$ and observe that

$$M = \{T \in \mathcal{I} : \exists \{n_j\} \text{ with } \lim_{j \to \infty} T^{n_j} x_l = \lambda x_l \; \forall l \in \mathbb{N}\}.$$

We see that $M = \bigcap_{k=1}^{\infty} M_k$ for the sets

$$M_k := \left\{ T \in \mathcal{I} : \sum_{l=1}^{\infty} \frac{1}{2^l \|x_l\|} \|T^n x_l - \lambda x_l\| < \frac{1}{k} \text{ for some } n \right\}$$

which are open for the strong operator topology. Therefore M is a G_δ-set which is dense by Lemma 3.10, and the residuality of M follows.

The unitary case goes analogously, and we now prove the more delicate contraction case. To do so we first show that

$$M = \{T \in \mathcal{C} : \exists \{n_j\}_{j=1}^{\infty} \text{ with } \lim_{j \to \infty} \langle T^{n_j} x_l, x_l \rangle = \lambda \|x_l\|^2 \; \forall l \in \mathbb{N}\}.$$

As in the proof of Theorem 3.9, the non-trivial inclusion follows from

$$\|(T^{n_j} - \lambda I)x\|^2 = \|T^{n_j} x\|^2 - 2\operatorname{Re} \langle T^{n_j} x, \lambda x \rangle + \|x\|^2$$
$$\leq 2\operatorname{Re} \left(\|x\|^2 - \langle T^{n_j} x, \lambda x \rangle \right) = 2\operatorname{Re} \left(\bar{\lambda} \langle (\lambda I - T^{n_j})x, x \rangle \right).$$

For the sets

$$M_k := \left\{ T \in \mathcal{C} : \sum_{l=1}^{\infty} \frac{1}{2^l \|x_l\|^2} |\langle (T^n - \lambda I)x_l, x_l \rangle| < \frac{1}{k} \text{ for some } n \right\},$$

we have the equality $M = \bigcap_{k=1}^{\infty} M_k$. Note again that it is not clear whether the sets M_k are open for the weak operator topology, so we use another argument to show that the complements M_k^c are nowhere dense. By Lemma 3.10 it suffices to show that $M_k^c \cap U_\lambda = \emptyset$ for the complement M_k^c and the set U_λ of all unitary operators U satisfying $U^n = \lambda I$ for some $n \in \mathbb{N}$. This can be shown as in the proof of Theorem 3.9 by replacing I by λI.

Since λ-rigidity for an irrational λ implies Γ-rigidity, the proof is complete. $\qquad \square$

We now present basic constructions leading to examples of operators with properties described in Theorem 3.11.

Example 3.12. a) A large class of abstract examples of Γ-rigid unitary operators which are almost weakly stable comes from harmonic analysis, more precisely, from Γ-rigid measures, see Subsection 1.1. Recall that λ-rigid continuous measures form a dense G_δ (and hence a residual) set in the space of all probability measures with respect to the weak* topology, see Theorem 1.14.

Take a λ-rigid measure μ for some irrational λ. The unitary operator given by $(Uf)(z) := zf(z)$ on $L^2(\Gamma, \mu)$ satisfies conditions (1) and (2) of Theorem 3.11. To show (2), it again suffices to prove that U is λ-rigid since λ is irrational. By our assumption on μ, there exists a subsequence $\{n_j\}_{j=1}^{\infty}$ such that the Fourier coefficients of μ satisfy $\lim_{j \to \infty} \hat{\mu}_{n_j} = \lim_{j \to \infty} \int_0^{2\pi} e^{in_j s} d\mu(s) = \lambda$, where we again identify Γ with $[0, 2\pi]$. Thus, $\lim_{j \to \infty} \langle U^{n_j} f, f \rangle = \lambda \|f\|^2$ for every $f \in L^2(\Gamma, \mu)$ with absolute value 1. By

$$\|U^{n_j} f - \lambda f\|^2 = \|U^{n_j} f\|^2 - 2\mathrm{Re}\left(\bar{\lambda} \langle U^{n_j} f, f \rangle\right) + \|f\|^2 = 2(\|f\|^2 - \mathrm{Re}\left(\bar{\lambda} \langle U^{n_j} f, f \rangle\right))$$

we see that $\lim_{j \to \infty} U^{n_j} f = \lambda f$ for every character, and therefore for every $f \in L^2(\Gamma, \mu)$ by the density argument, implying that U is Γ-rigid.

Generally, a unitary operator U on a separable Hilbert space satisfies conditions (1) and (2) of Theorem 3.11, i.e., is Γ-rigid and almost weakly stable if and only if every spectral measure of U is continuous and Γ-rigid. The "if" direction follows by arguments as in the proof of Proposition 3.3.

b) Another class of examples of rigid almost weakly stable unitary operators comes from ergodic theory. A measure preserving transformation φ on a probability space (Ω, Σ, μ) is called *rigid* if there exists a subsequence $\{n_j\}_{j=1}^{\infty}$ such that

$$\lim_{j \to \infty} \mu(A \triangle \varphi^{-n_j}(A)) = 0 \quad \text{for every } A \in \Sigma.$$

Consider now the induced operator T on $H := L^2(\Omega, \mu)$ defined by $(Tf)(\omega) := f(\varphi(\omega))$. By

$$\mu(A \triangle \varphi^{-n_j}(A)) = \int_\Omega |\mathbf{1}_{\varphi^{-n_j}(A)} - \mathbf{1}_A|^2 d\mu = \|T^{n_j} \mathbf{1}_A - \mathbf{1}_A\|^2$$

and since characteristic functions span $L^2(\Omega, \mu)$, we see that T is rigid if and only if the transformation φ is rigid. Thus, restricting T to the invariant subspace $H_0 := \{f : \int_\Omega f d\mu = 0\}$, we obtain that every rigid weakly mixing transformation induces a rigid almost weakly stable unitary operator, cf. Corollary 2.7. Analogously, a transformation φ is called λ-rigid (or λ-weakly mixing) if the restriction T_0 of T to H_0 is λ-rigid. Thus, each weakly mixing λ-rigid transformation leads to a unitary operator satisfying conditions (1) and (2) in Theorem 3.11.

Katok [142] proved that rigid transformations form a dense G_δ-set in the set of all measure preserving transformations, and Choksi, Nadkarni [51] generalised this to λ-rigid transformations. For more information we refer to Nadkarni [193, p. 59] and for concrete examples of rigid weakly mixing transformations using adding machines and interval exchange transformations see Goodson, Kwiatkowski, Lemańczyk, Liardet [109] and Ferenczi, Holton, Zamboni [83], respectively. For examples of rigid weakly mixing transformations given by Gaussian automorphisms see Cornfeld, Fomin, Sinai [53, Chapter 14]. Finally, a whole class of examples of rigid weakly mixing transformations comes from special flows, see Lemanczyk, Mauduit [165].

Furthermore, there is an (abstract) method of constructing λ-rigid operators from a rigid one. The idea of this construction in the context of measures belongs to Nadkarni [193, Chapter 7].

Example 3.13. Let T be a rigid contraction with $\lim_{j\to\infty} T^{n_j} = I$ strongly. We construct a class of λ-rigid operators from T. Note that if λ is irrational and if T is unitary with no point spectrum, this construction gives us a class of examples satisfying (1)–(2) of Theorem 3.11.

Take $\alpha \in \Gamma$ and consider the operator $T_\alpha := \alpha T$. Then we see that T_α is λ-rigid if $\lim_{j\to\infty} \alpha^{n_j} = \lambda$ for the above sequence $\{n_j\}$. Nadkarni [193, pp. 49–50] showed that the set of all α such that the limit set of $\{\alpha^{n_j}\}_{j=1}^\infty$ contains an irrational number has full Lebesgue measure in Γ. Every such α leads to a Γ-rigid operator αT.

We now show that one cannot replace the operators λI in Theorem 3.11 by any other operator.

Proposition 3.14. *Let $V \in \mathcal{L}(H)$ be such that the set*

$$M_V := \{T : \exists \{n_j\}_{j=1}^\infty \text{ such that } \lim_{j\to\infty} T^{n_j} = V \text{ strongly}\}$$

is dense in one of the spaces $\mathcal{U}, \mathcal{I}$ or $\mathcal{C}$. Then V is a multiple of identity.

Proof. Consider the contraction case and assume that the set of all contractions T such that weak-$\lim_{j\to\infty} T^{n_j} = V$ for some $\{n_j\}_{j=1}^\infty$ is dense in $\mathcal{C}$. Since every such operator T commutes with V by $TV = \lim_{j\to\infty} T^{n_j+1} = VT$, we obtain by assumption that V commutes with every contraction. In particular, V commutes with every one-dimensional projection implying that $V = \lambda I$ for some $\lambda \in \mathbb{C}$.

The same argument works for the spaces $\mathcal{I}$ and $\mathcal{U}$ using the density of unitary operators in the set of all contractions for the weak operator topology. $\square$

Remark 3.15. In the above proposition, one has $V = \lambda I$ for some $\lambda \in \Gamma$ in the unitary and isometric case. Moreover, the same holds in the contraction case if M_V is residual. (This follows from the fact that the set of all non-unitary contractions is of first category in $\mathcal{C}$ by Remark 3.2 and Theorem 3.9, see also Theorem V.1.22 below.

Remark 3.16. It is not clear whether Theorem 3.11 remains valid under the additional requirement

$$\{\lambda \in \mathbb{C} : |\lambda| < 1\} \cdot I \subset \overline{\{T^n : n \in \mathbb{N}\}}^{\sigma}, \tag{IV.6}$$

where σ denotes the weak operator topology. Since countable intersections of residual sets are residual and the right-hand side of (IV.6) is closed, this question takes the following form: Is, for a fixed λ with $0 < |\lambda| < 1$, the set M_λ of all contractions T satisfying $\lambda I \in \overline{\{T^n : n \in \mathbb{N}\}}^{\sigma}$ residual? Note that each M_λ is dense in $\mathcal{C}$ since, for $\lambda = re^{is}$, it contains the set

$$\{cU : 0 < c < 1, \ c^n = r \text{ and } U^n = e^{is}I \text{ for some } n\}$$

which is dense in $\mathcal{C}$ by the density of unitary operators and a natural modification of Lemma 3.10.

We finally mention that absence of rigidity does not imply weak stability, or, equivalently, absence of weak stability does not imply rigidity, as shown by the following example.

Example 3.17. There exist unitary operators T with no non-trivial weakly stable orbit which are nowhere rigid, i.e., such that $\lim_{j\to\infty} T^{n_j}x = x$ for some subsequence $\{n_j\}_{j=1}^{\infty}$ implies $x = 0$. (Note that such operators are automatically almost weakly stable by Proposition 3.3 and the Jacobs–Glicksberg–de Leeuw decomposition.) A class of such examples comes from *mildly mixing* transformations which are not strongly mixing, see Furstenberg, Weiss [93], Fraçzek, Lemańczyk [90] and Fraçzek, Lemańczyk, Lesigne [91].

3.3 Continuous case: C_0-semigroups

We now give the continuous analogue of the above results for unitary and isometric strongly continuous (semi)groups.

Let H be again a separable infinite-dimensional Hilbert space. We recall the objects introduced in Section III.6. We denote by $\mathcal{U}^{\text{cont}}$ the set of all unitary C_0-groups on H endowed with the topology of strong convergence of semigroups and their adjoints uniformly on compact time intervals. This is a complete metric and hence a Baire space for

$$d(U(\cdot), V(\cdot)) := \sum_{n,j=1}^{\infty} \frac{\sup_{t\in[-n,n]} \|U(t)x_j - V(t)x_j\|}{2^j \|x_j\|} \quad \text{for } U(\cdot), V(\cdot) \in \mathcal{U}^{\text{cont}},$$

where $\{x_j\}_{j=1}^{\infty}$ is a fixed dense subset of $H \setminus \{0\}$. We further denote by $\mathcal{I}^{\text{cont}}$ the set of all isometric C_0-semigroups on H endowed with the topology of strong convergence uniform on compact time intervals. Again, this is a complete metric space for

$$d(T(\cdot), S(\cdot)) := \sum_{n,j=1}^{\infty} \frac{\sup_{t\in[0,n]} \|T(t)x_j - S(t)x_j\|}{2^j \|x_j\|} \quad \text{for } T(\cdot), S(\cdot) \in \mathcal{I}^{\text{cont}}.$$

The proofs of the following results are similar to the discrete case, but require some additional technical details.

Theorem 3.18. *Let H be a separable infinite-dimensional Hilbert space. The set*

$$M^{\text{cont}} := \{T(\cdot): \lim_{j\to\infty} T(t_j) = I \text{ strongly for some } t_j \to \infty\}$$

is residual in the set $\mathcal{I}^{\text{cont}}$ of all isometric C_0-semigroups for the topology corresponding to strong convergence uniform on compact time intervals in $\mathbb{R}_+$. The same holds for unitary C_0-groups and the topology of strong convergence uniform on compact time intervals in $\mathbb{R}$.

Proof. We begin with the unitary case.

Choose $\{x_l\}_{l=1}^{\infty}$ as a dense subset of $H \setminus \{0\}$. Since one can replace $\lim_{j\to\infty} t_j = \infty$ in the definition of M^{cont} by $\{t_j\}_{j=1}^{\infty} \subset [1, \infty)$ we have

$$M^{\text{cont}} = \{T(\cdot) \in \mathcal{U}^{\text{cont}} : \exists \{t_j\} \in [1, \infty) : \lim_{j\to\infty} T(t_j)x_l = x_l \; \forall l \in \mathbb{N}\}. \quad (\text{IV.7})$$

Consider now the open sets

$$M_{k,t} := \left\{ T(\cdot) \in \mathcal{U}^{\text{cont}} : \sum_{l=1}^{\infty} \frac{1}{2^l \|x_l\|} \|T(t)x_l - x_l\| < \frac{1}{k} \right\}$$

and $M_k^{\text{cont}} := \bigcup_{t\geq 1} M_{k,t}$. We have

$$M^{\text{cont}} = \bigcap_{k=1}^{\infty} M_k^{\text{cont}},$$

and hence M^{cont} is a G_δ-set. Since periodic unitary C_0-groups are dense in $\mathcal{U}^{\text{cont}}$ by Proposition III.6.1, and since they are contained in M^{cont}, we see that M^{cont} is residual as a countable intersection of dense open sets.

The same arguments and the density of periodic unitary operators in $\mathcal{I}^{\text{cont}}$ (see Proposition III.6.7) imply the assertion in the isometric case. $\qquad\square$

The following continuous analogue of Lemma 3.10 allows us to replace I by λI.

Lemma 3.19. *Let H be a Hilbert space and fix $\lambda \in \Gamma$ and $N \in \mathbb{N}$. Then for every unitary C_0-group $U(\cdot)$ there exists a sequence $\{U_n(\cdot)\}_{n=1}^{\infty}$ of unitary C_0-groups such that*

(a) *For every $n \in \mathbb{N}$ there exists $\tau \geq N$ with $U_n(\tau) = \lambda I$,*

(b) $\lim_{n \to \infty} \|U_n(t) - U(t)\| = 0$ *uniformly on compact intervals in $\mathbb{R}$.*

Proof. Let $U(\cdot)$ be a unitary C_0-group on H, $\lambda = e^{i\alpha} \in \Gamma$. By the spectral theorem, H is isomorphic to $L^2(\Omega, \mu)$ for some finite measure space (Ω, μ) and $U(\cdot)$ is unitarily equivalent to a multiplication group $\tilde{U}(\cdot)$ given by

$$(\tilde{U}(t)f)(\omega) = e^{itq(\omega)}f(\omega), \quad \omega \in \Omega,$$

for some measurable $q : \Omega \to \mathbb{R}$.

To approximate $\tilde{U}(\cdot)$, let $N \in \mathbb{N}$, $\varepsilon > 0$, $t_0 > 0$ and take $m \geq N$, $m \in \mathbb{N}$, such that $\|1 - e^{\frac{2\pi i}{m}}\| \leq \varepsilon/(2t_0)$. Define for $\alpha_j := e^{i\left(\frac{\alpha}{m} + \frac{2\pi j}{m}\right)}$, $j = 0, \ldots, m$,

$$p(\omega) := \alpha_{j-1} \text{ for all } \omega \in \varphi^{-1}(\{z \in \Gamma : \arg(\alpha_{j-1}) \leq \arg(z) < \arg(\alpha_j)\}).$$

The multiplication group $\tilde{V}(\cdot)$ defined by $\tilde{V}(t)f(\omega) := e^{itp(\omega)f(\omega)}$ satisfies $\tilde{V}(m) = e^{i\alpha}$. Moreover,

$$\|\tilde{U}(t)f - \tilde{V}(t)f\|^2 = \int_{\Omega} |e^{itq(\omega)} - e^{itp(\omega)}|^2 \|f(\omega)\|^2$$
$$\leq 2|t| \sup_{\omega \in \Omega} |q(\omega) - p(\omega)| \|f\|^2 < \varepsilon \|f\|^2$$

uniformly in $t \in [-t_0, t_0]$. $\qquad\square$

We now obtain the following characterisation of the "typical" asymptotic behaviour of isometric and unitary C_0-(semi)groups on separable Hilbert spaces.

Theorem 3.20. *Let H be a separable infinite-dimensional Hilbert space. Then the set of all C_0-semigroups $T(\cdot)$ on H satisfying the following properties*

(1) *there exists a set $M \subset \mathbb{R}_+$ with density 1 such that*

$$\lim_{t \to \infty, t \in M} T(t) = 0 \quad \text{weakly,}$$

(2) *for every $\lambda \in \Gamma$ there exists $\{t_j^{(\lambda)}\}_{j=1}^{\infty}$ with $\lim_{j \to \infty} t_j^{(\lambda)} = \infty$ such that*

$$\lim_{j \to \infty} T(t_j^{(\lambda)}) = \lambda I \quad \text{strongly}$$

is residual in the set of all isometric C_0-semigroups for the topology of strong convergence uniform on compact time intervals in $\mathbb{R}_+$. The same holds for unitary C_0-groups for the topology of strong convergence uniform on compact time intervals in $\mathbb{R}$.

Proof. By Theorems III.6.5 and III.6.9, C_0-(semi)groups satisfying (1) are residual in $\mathcal{I}^{\text{cont}}$ and $\mathcal{U}^{\text{cont}}$.

We show, for a fixed $\lambda \in \Gamma$, the residuality of the set $M^{(\lambda)}$ of all C_0-semigroups $T(\cdot)$ satisfying strong-$\lim_{j\to\infty} T(t_j) = \lambda I$ for some sequence $\{t_j\}_{j=1}^{\infty}$ converging to infinity. We prove this property for the space $\mathcal{I}^{\text{cont}}$ of all isometric semigroups and the strong operator convergence uniform on compact intervals, the unitary case goes analogously.

Take $\lambda \in \Gamma$ and observe

$$M^{(\lambda)} = \{T(\cdot) \in \mathcal{I}^{\text{cont}} : \exists \{t_j\} \subset [1, \infty) \text{ with } \lim_{j\to\infty} T(t_j)x_l = \lambda x_l \; \forall l \in \mathbb{N}\}$$

for a fixed dense sequence $\{x_l\}_{l=1}^{\infty} \subset H \setminus \{0\}$. Consider now the open sets

$$M_{k,t} := \left\{ T(\cdot) \in \mathcal{I}^{\text{cont}} : \sum_{l=1}^{\infty} \frac{\|(T(t) - \lambda I)x_l\|}{2^l \|x_l\|} < \frac{1}{k} \right\}$$

and their union $M_k := \bigcup_{t \geq 1} M_{k,t}$ being open as well. The equality $M^{(\lambda)} = \bigcap_{k=1}^{\infty} M_k$ follows as in the proof of Theorem 3.18. Since every M_k is dense by Lemma 3.19, $M^{(\lambda)}$ is residual as a dense countable intersection of open sets.

Since λ-rigidity for some irrational λ implies λ-rigidity for every $\lambda \in \Gamma$, the theorem is proved. $\square$

Note that every semigroup satisfying (1) and (2) above is a unitary group by Remark 3.7.

Example 3.21. a) There is the same correspondence between rigid (or λ-rigid) unitary C_0-groups and rigid (or λ-rigid) probability measures on $\mathbb{R}$ as in the discrete case, see Example 3.12. More precisely, to a probability measure μ on $\mathbb{R}$ one associates the multiplication group given by $(T(t))f(s) := e^{ist}f(s)$ on $H = L^2(\mathbb{R}, \mu)$. For $\lambda \in \Gamma$, we call a measure μ on $\mathbb{R}$ λ-*rigid* if there exists $t_j \to \infty$, $t_j \in \mathbb{R}$, such that the Fourier transform of μ satisfies $\lim_{j\to\infty} \mathcal{F}\mu(t_j) = \lambda$. By Theorem 1.16 the set of all continuous Γ-rigid measures on $\mathbb{R}$ is a residual (in fact a dense G_δ) set in the set of all Radon measures with respect to the weak* topology. For each such measure, the associated unitary C_0-group is Γ-rigid and almost weakly stable, and conversely, the spectral measures of an almost weakly stable Γ-rigid unitary group are continuous and Γ-rigid.

b) Again, another large class of examples comes from ergodic theory. Consider a measure preserving semiflow $\{\varphi_t\}_{t \geq 0}$ on a probability space (Ω, μ). The induced semigroup $T(\cdot)$ on $L^2(\Omega, \mu)$ is almost weakly stable and λ-rigid if and only if the semiflow (φ_t) is weakly mixing and λ-rigid, where the last notion is defined analogously to the discrete case. However, it seems that there is no result in ergodic theory stating that a "typical" (semi)flow is rigid and weakly mixing. So we apply the following abstract argument to present a large class of such flows.

We start from a probability space (Ω, μ). As discussed above, a "typical" measure preserving transformation φ on (Ω, μ) is weakly mixing and Γ-rigid. On the other hand, by de la Rue, de Sam Lazaro [59] a "typical" (for the same topology) measure preserving transformation φ is embeddable into a flow, i.e., there exists a flow $(\varphi_t)_{t\in\mathbb{R}}$ such that $\varphi = \varphi_1$. Therefore, a "typical" transformation is weakly mixing, Γ-rigid *and* embeddable, and every such transformation leads to an almost weakly stable Γ-rigid unitary group.

For examples of weakly mixing special flows which are rigid see, e.g., Lemanczyk, Mauduit [165].

c) Analogously to b), we can construct a more general class of examples on Hilbert spaces. We use that every unitary operator T is embeddable into a unitary C_0-group $T(\cdot)$, see Corollary V.1.15. Take now any operator satisfying assertions of Theorem 3.11. Since such an operator is automatically unitary by Remark 3.2, it is embeddable. Thus, every such operator leads to an example of a C_0-group satisfying (1) and (2) of Theorem 3.20. (Note that condition (1) follows from the spectral mapping theorem for the point spectrum, see e.g. Engel, Nagel [78, Theorem IV.3.7].)

We again show that the limit operators λI, $|\lambda| = 1$, cannot be replaced by any other operator.

Proposition 3.22. *Let for some $V \in \mathcal{L}(H)$ the set*

$$M_V^{cont} := \{T(\cdot) : \exists t_j \to \infty \text{ such that } \lim_{j\to\infty} T(t_j) = V \text{ strongly}\}$$

be dense in $\mathcal{I}^{cont}$ or $\mathcal{U}^{cont}$. Then $V = \lambda I$ for some $\lambda \in \Gamma$.

Proof. We prove this assertion for $\mathcal{I}$, the unitary case is analogous.

Observe that V commutes with every $T(\cdot) \in M_V^{cont}$ by

$$VT(t) = \text{strong-}\lim_{j\to\infty} T(t_j + t) = T(t)V$$

implying by assumption that V commutes with every unitary C_0-group. Since unitary C_0-groups are dense in the set of all contractive C_0-semigroups for the topology of weak operator convergence uniform on compact time intervals, proven by Król [155], we see that V commutes with every contractive C_0-semigroup. Since orthogonal one-dimensional projections are embeddable into a contractive C_0-semigroup by Proposition V.1.13, V commutes with every orthogonal one-dimensional projection, which implies $V = \lambda I$ for some $\lambda \in \mathbb{C}$. Moreover, $|\lambda| = 1$ holds since V is the strong limit of isometric operators. $\qquad\Box$

Remark 3.23. It is again not clear whether one can formulate an analogue of the above result for contractive C_0-semigroups as done in the discrete case, cf. Subsection III.6.3.

3.4 Further remarks

We now consider some generalisations of the above results.

"Controlling" the sequences $\{n_j\}$ and $\{t_j\}$

We take a closer look at the sequences $\{n_j\}$ and $\{t_j\}$ occuring in Theorems 3.11(2) and 3.20(2).

Observe first, by the same arguments as in the proofs of Theorems 3.9 and 3.11, that we can replace T by T^m for a fixed m. Changing appropriately the assertion and the proof of Lemma 3.10, we see that one can add the condition $\{n_j\}_{j=1}^{\infty} \subset m\mathbb{N}$ to the sequence yielding rigidity and λ-rigidity. More precisely, for every $\lambda \in \Gamma$ and $m \in \mathbb{N}$, the set of all operators T such that strong-$\lim_{j\to\infty} T^{n_j} = I$ for some $\{n_j\} \subset m\mathbb{N}$ is residual in $\mathcal{U}$, $\mathcal{I}$ and $\mathcal{C}$, and analogously for λI.

It is a hard problem to determine the sequences $\{n_j\}$ and $\{t_j\}$ exactly. However, one can generalise the above observation and "control" these sequences in the following sense. Let $\Lambda \subset \mathbb{N}$ be an unbounded set. We call an operator T *rigid along* Λ if strong-$\lim_{j\to\infty} T^{n_j} = I$ for some increasing sequence $\{n_j\}_{j=1}^{\infty} \subset \Lambda$. Similarly, we define rigidity along an unbounded set $\Lambda \subset \mathbb{R}_+$ for C_0-semigroups, as well as λ- and Γ-rigidity. It follows from a natural modification of Lemmas 3.10 and 3.19 that, for a fixed unbounded set Λ in $\mathbb{N}$ and $\mathbb{R}_+$, respectively, one can assume $\{n_j^{(\lambda)}\} \subset \Lambda$ and $\{t_j^{(\lambda)}\} \subset \Lambda$ in Theorems 3.11(2) and 3.20(2). Thus, the set of all Γ-rigid operators (semigroups) along a fixed unbounded set is residual in $\mathcal{U}$, $\mathcal{I}$ and $\mathcal{C}$ ($\mathcal{U}^{\text{cont}}$ and $\mathcal{I}^{\text{cont}}$, respectively).

Banach space case

We finally discuss briefly the situation in Banach spaces.

Note first that Theorems 3.11 and 3.20 are not true in general separable Banach spaces. Indeed, since weak convergence in l^1 implies strong convergence, we see that (2) implies strong convergence to zero of T^n (or of $T(t)$, respectively), making (3) or just rigidity impossible.

We now consider the question in which Banach spaces rigid and Γ-rigid operators are residual. Since in the contraction case our techniques heavily use Hilbert space methods, we only consider the isometric and unitary case. Let X be a separable infinite-dimensional Banach space, and $\mathcal{I}$ be the set of all isometries on X endowed with the strong operator topology. Observe that the sets

$$M_k := \{T \in \mathcal{I} : \sum_{l=1}^{\infty} \frac{1}{2^l \|x_l\|} \|(T^n - I)x_l\| < \frac{1}{k} \text{ for some } n\}$$

appearing in the proof of Theorem 3.9 for the isometric case are still open, and therefore M is a G_δ-set containing periodic isometries. Thus Theorem 3.9 holds in all separable infinite-dimensional Banach spaces such that periodic isometries form a dense set of $\mathcal{I}$. Analogously, the set of operators satisfying (1) and (2) of

Theorem 3.11 is residual in $\mathcal{I}$ if and only if it is dense in $\mathcal{I}$. The same assertions hold for the set $\mathcal{U}$ of all invertible isometric operators with the topology induced by the seminorms $p_x(T) = \sqrt{\|Tx\|^2 + \|T^{-1}x\|^2}$, which is a complete metric space with respect to the metric

$$d(T,S) = \sum_{j=1}^{\infty} \frac{\|Tx_j - Sx_j\| + \|T^{-1}x_j - S^{-1}x_j\|}{2^j \|x_j\|}$$

for a fixed dense sequence $\{x_j\}_{j=1}^{\infty} \subset X \setminus \{0\}$.

Chapter V

Discrete vs. continuous

Having investigated discrete and continuous systems $\{T^n\}_{n=0}^{\infty}$ and $(T(t))_{t\geq 0}$ separately, we now try to "build bridges" between them.

In the first section we embed discrete systems into continuous ones, a delicate problem to which we can give some partial, but new, answers. In the second part we associate to each continuous system a bounded operator, the cogenerator, introduced in Section I.3. We show that many asymptotic properties of the continuous system are reflected by analogous properties of the discrete powers of this cogenerator.

1 Embedding operators into C_0-semigroups

It is an old philosophical question whether time should be considered to be discrete or continuous. So, it is very natural to ask which discrete dynamical systems come from continuous ones. More precisely, we study the following property.

Definition 1.1. We will say that a linear operator T on a Banach space X can be embedded into a C_0-semigroup (shortly, is *embeddable*) if there exists a C_0-semigroup $(T(t))_{t\geq 0}$ on X such that $T = T(1)$.

In other words, we are interested in operators which appear as an element of a C_0-semigroup.

Note that the embedding property implies the existence of roots of all orders of T. Furthermore, T and $(T(t))_{t\geq 0}$ share properties such as norm/strong/weak/almost weak stability by Theorem III.2.2, Lemma III.3.4, Theorem III.4.4, the spectral mapping theorem for the point spectrum and Remark I.1.18, respectively. Clearly, the semigroup $T(\cdot)$ is not unique since, e.g., all semigroups $(T_n(t)) := (e^{2\pi i n t}T(t))$ again satisfy $T_n(1) = T(1)$. We will concentrate on the problem to find a C_0-semigroup into which a given operator can be embedded.

The question is difficult and has analogues in other areas of mathematics like ergodic theory (see e.g. King [149], de la Rue, de Sam Lasaro [59], Stepin,

Eremenko [235]), stochastics and measure theory (see e.g. Heyer [128, Chapter III] and Fischer [84]). As a first answer we discuss how spectral calculus leads to a sufficient condition for embeddability which, however, is far from being necessary. We also present a completely different, but simple, necessary condition which is, for some classes of operators, also sufficient. Finally, we show that a "typical" (in the sense of Baire category) contraction on a separable Hilbert space is embeddable. We follow [66].

1.1 Sufficient conditions via functional calculus

We start with the classical approach using functional calculus. It is based on a sufficient spectral condition allowing us to construct the generator of $T(\cdot)$ as a logarithm of T.

Theorem 1.2. *Assume that $\sigma(T)$ is contained in a simply connected open domain which does not include $\{0\}$. Then T can be embedded into a uniformly continuous C_0-semigroup.*

Proof. By the Dunford functional calculus, we can define $A := \ln(T)$ as a bounded operator. Then we have $T = e^A$ and T can be embedded into the semigroup $(T(t))_{t\geq 0} = (e^{tA})_{t\geq 0}$ which is uniformly continuous. $\qquad\square$

For an abstract version of this result and basic properties of the exponential function and the logarithm in the context of unital Banach algebras see Palmer [208, Theorems 2.1.12 and 3.4.4].

Remark 1.3. Note that one can construct $\ln T$ as a bounded operator also in some more cases. For example, if X is a UMD-space and the Cayley transform $A := (I + T)(I - T)^{-1}$ exists and generates a C_0-group with exponential growth < 1, then $\ln(T)$ exists and is bounded, and hence T can be embedded into a uniformly continuous C_0-semigroup. For details see Haase [115, Example 3.7].

There are several extensions of Theorem 1.2. One is the following result allowing us to construct semigroups with unbounded generators. Recall that an operator is called *sectorial* if there exists a sector $\Sigma_\delta = \{z : |\arg z| \leq \delta\} \cup \{0\}$, $0 < \delta < \pi$, such that $\sigma(T) \subset \Sigma_\delta$ and for every $\omega > \delta$ one has $\|R(\lambda, T)\| \leq \frac{M}{|\lambda|}$ for some M and every $\lambda \notin \Sigma_\omega$.

Theorem 1.4. (See Haase [114, Prop. 3.1.1 and 3.1.15]) *Let T be a bounded sectorial operator with dense range. Then T can be embedded into an analytic C_0-semigroup.*

Since T is embeddable if and only if cT is embeddable for any $0 \neq c \in \mathbb{C}$ (consider $T_c(t) := r^t e^{i\varphi t} T(t)$ for $c = re^{i\varphi}$), one can also formulate an analogous result for operators with spectrum in a rotated sector $\{z : |\arg z - \varphi| < \delta\}$.

We now show that the spectrum of an embeddable operator can be arbitrary, hence conditions on the location of the spectrum are not necessary and the functional calculus method works only in particular cases.

Example 1.5. Let $K \neq \emptyset$ be an arbitrary closed subset in $\mathbb{C}$. Assume first that 0 is not an isolated point of K and take a dense subset $\{\lambda_k\}_{k=1}^\infty$ of $K \setminus \{0\}$. Consider the space $X := l^2$ and the multiplication operator T given by

$$T(x_1, x_2, \ldots) := (\lambda_1 x_1, \lambda_2 x_2, \ldots)$$

on X with $\sigma(T) = K$. Define now the semigroup $(T(t))_{t \geq 0}$ by

$$T(t)(x_1, x_2, \ldots) := (e^{t \ln \lambda_1} x_1, e^{t \ln \lambda_2} x_2, \ldots),$$

where we take an arbitrary value of the logarithm of each λ_k. The family $T(\cdot)$ forms a semigroup of contractions which is strongly continuous since $\lim_{t \to 0+} T(t) e_k = e_k$ for every basis vector e_k and $T(\cdot)$ is uniformly bounded on compact time intervals. Assume now that 0 is an isolated point of K, i.e., $K = \{0\} \cup K_1$ for a compact set K_1 with $0 \notin K_1$, and define T_1 with $\sigma(T_1) = K_1$ as above. Define further $T_2 := 0$ on l^2 which is embeddable into a nilpotent semigroup by Lemma 1.12 below. Thus the direct sum of T_1 and T_2 is embeddable with spectrum equal to K.

Remark 1.6. Using an analogous method, one can construct for every compact set K a non-embeddable operator T with $\sigma(T) = K \cup \{0\}$: Take just the direct sum of the operator constructed in the first part of Example 1.5 with the zero operator on the one-dimensional space. This sum is not embeddable by Theorem 1.7 below.

1.2 A necessary condition

In this section we present a (simple) necessary condition for embeddability using information on the spectral value 0.

Theorem 1.7. *Let X be a Banach space and $T \in \mathcal{L}(X)$. If T can be embedded into a C_0-semigroup, then $\dim(\ker T)$ and $\operatorname{codim}(\overline{\operatorname{rg} T})$ are zero or infinite.*

In other words, operators with $0 < \dim(\ker T) < \infty$ or $0 < \operatorname{codim}(\overline{\operatorname{rg} T}) < \infty$ cannot be embedded into a C_0-semigroup.

Proof. Assume that $0 < \dim(\ker T) < \infty$ holds and $T = T(1)$ for some C_0-semigroup $T(\cdot)$ on X. Since T is not injective, neither is its square root $T(\frac{1}{2})$. Analogously, every $T(\frac{1}{2^n})$ is not injective. Take $x_n \in \ker T(\frac{1}{2^n})$ with $\|x_n\| = 1$ for every $n \in \mathbb{N}$. Then we have $\{x_n\}_{n=1}^\infty \subset \ker T$ and, since $\ker T$ is finite-dimensional, there exists a subsequence $\{x_{n_k}\}_{k=1}^\infty$ converging to some x_0 with $\|x_0\| = 1$.

Since $\ker T(\frac{1}{2^{n+1}}) \subset \ker T(\frac{1}{2^n})$, we have $T(\frac{1}{2^n}) x_0 = 0$ for every $n \in \mathbb{N}$ contradicting the strong continuity of $T(\cdot)$.

Assume now that $0 < \operatorname{codim}(\overline{\operatorname{rg} T}) = \dim(\ker T') < \infty$ holds. If T is embedded into a C_0-semigroup $T(\cdot)$, then T' embeds into the adjoint semigroup $T'(\cdot)$ on X' which is weak* continuous. The same arguments as above contradict the weak* continuity. $\square$

A direct corollary is the following.

Corollary 1.8. *Non-bijective Fredholm operators are not embeddable.*

Remark 1.9. As we will see in Subsection 1.4, all four possibilities given in Theorem 1.7 (dimension 0 or ∞ of the kernel and codimension 0 or ∞ of the closure of the range) can appear for operators having the embedding property. The examples are provided by unitary operators, the left and right shifts on $l^2(Y)$ for an infinite-dimensional Hilbert space Y and their direct sum, respectively. Note moreover that, by Theorems 1.7 and 1.2, an operator on a finite-dimensional space is embeddable if and only if its spectrum does not contain 0.

In the following we ask the converse question.

Question 1.10. For which (classes of) operators is the necessary condition given in Theorem 1.7 also sufficient for embeddability?

Note that this condition is not sufficient for embeddability in general. For example, Halmos, Lumer and Schäffer [122] constructed a class of *invertible* operators on Hilbert spaces having no square root and hence not being embeddable. Later, Deckard and Pearcy [58] presented another class of such operators using a matrix construction. We present here the example by Halmos, Lumer and Schäffer [122].

Example 1.11. Consider $D := \{z \in \mathbb{C} : 1 < |z| < 2\}$ and the corresponding Bergman space H consisting of all holomorphic functions $f : D \to \mathbb{C}$ satisfying $\|f\|_2^2 := \int_D |f(z)|^2 \, d\mu < \infty$ for the planar Lebesgue measure μ. Then H is a Hilbert space with the standard scalar product $\langle f, g \rangle = \int_D f(z)\overline{g(z)} \, d\mu$. (For basic properties of Bergman spaces see Hedenmalm, Korenblum, Zhu [124] or Bergman [32].) Consider further the multiplication operator T given by $(Tf)(z) := zf(z)$. Then $\sigma(T) = \overline{D}$, and hence T is invertible.

We now need more detailed information on the spectrum of T. Take $\lambda \in D$. We first show that

$$f \in \mathrm{rg}(\lambda I - T) \quad \text{if and only if} \quad f(\lambda) = 0. \tag{V.1}$$

Indeed, $f = \lambda g - Tg$ implies $f(\lambda) = \lambda g(\lambda) - \lambda g(\lambda) = 0$. Conversely, if $f(\lambda) = 0$, then by the local power series representation of f we see that $g(z) := \frac{f(z)}{\lambda - z}$ is holomorphic in D. To show that $g \in H$, i.e., $\|g\|_2 < \infty$, let $\delta := \frac{\mathrm{dist}(\lambda, \partial D)}{2}$, where ∂D denotes the boundary of D. On $\overline{U_\delta(\lambda)}$ the function g is bounded, hence square integrable. On the other hand, g is dominated by f/δ on $D \setminus U_\delta(\lambda)$ and therefore square integrable as well. This implies $g \in H$ and thus $f \in \mathrm{rg}(\lambda I - T)$.

It is well-known that the point evaluation $\varphi(f) := f(\lambda)$ is continuous on H, see e.g. Bergman [32, p. 24] or Halmos, Lumer and Schäffer [122]. We can reformulate (V.1) as $\mathrm{rg}(\lambda I - T) = \ker \varphi$. So the subspace $\ker(\lambda I - T') = \mathrm{rg}(\lambda I - T)^\perp = (\ker \varphi)^\perp$ is one-dimensional, and we obtain the following key property:

$$\text{every } \lambda \in D \text{ is a simple eigenvalue of } T'. \tag{V.2}$$

This implies in particular that $D \subset R_\sigma(T)$.

We now show that T has no square root. Observe first that $\sqrt{D} := \{z \in \mathbb{C} : z^2 \in D\} = \{z \in \mathbb{C} : 1 < |z| < \sqrt{2}\}$ is connected. Assume that there exists $S \in \mathcal{L}(X)$ with $S^2 = T$ and take $\lambda \in \sqrt{D}$. By the spectral mapping theorem for polynomials, λ or $-\lambda$ belong to $\sigma(S)$ and more precisely to $R_\sigma(S)$. Moreover, by (V.2) and the spectral mapping theorem again, only one of the numbers λ and $-\lambda$ can belong to $R_\sigma(S)$. Consequently, we obtain the decomposition

$$\sqrt{D} = (\sqrt{D} \cap \sigma(S)) \cup (\sqrt{D} \cap (-\sigma(S)))$$

into two closed subsets in the relative topology. This contradicts the connectedness of $\sqrt{D}$, and so T has no square root.

The same arguments show that T has no nth root for every $n \in \{2, 3, \dots\}$. Note further that the above construction works for any domain D with connected $\sqrt{D}$ or $\sqrt[n]{D}$, respectively.

1.3 Normal operators and projections on Hilbert spaces

It follows from the spectral theorem that every unitary operator can be embedded into a unitary group (cf. Corollary 1.15 below). We show that the condition in Theorem 1.7 suffices to embed normal operators and projections on Hilbert spaces. To do this we first need the following simple lemma.

Lemma 1.12. *Let H be an infinite-dimensional Hilbert space. Then the zero operator on H can be embedded into a C_0-semigroup.*

Proof. Observe that the Hilbert space H is unitarily isomorphic to $L^2([0,1], H)$. The assertion follows from the fact that the zero operator on $L^2([0,1], H)$ can be embedded into the nilpotent shift semigroup given by $T(t) = 0$ for $t \geq 1$ and

$$(T(t)f)(s) = \begin{cases} f(t-s), & s \in [0, t], \\ 0, & s \in (t, 1] \end{cases}$$

for $t \in [0, 1)$, $f \in L^2([0,1], H)$, $s \in [0, 1]$. $\qquad\square$

Embedding projections on Hilbert spaces is a direct corollary.

Proposition 1.13. *Let P be a projection on a Hilbert space H. Then P is embeddable if and only if $P = I$ or $\dim(\ker P) = \infty$. Moreover, in this case P can be embedded into a contractive C_0-semigroup if and only if P is an orthogonal projection.*

Proof. If $0 < \dim(\ker P) < \infty$, P is not embeddable by Theorem 1.7. The rest follows from the decomposition $H = \ker P \oplus \operatorname{rg} P$ (which is orthogonal if and only if P is orthogonal), Lemma 1.12 and embeddability of the identity operator (on any Banach space) into the identity semigroup. $\qquad\square$

We now characterise embeddable normal operators.

Theorem 1.14. *Let T be a normal operator on a Hilbert space. Then T is embeddable if and only if T is injective or $\dim \ker T = \infty$.*

Proof. By the spectral theorem we may assume that T is a multiplication operator on $H = L^2(\Omega, \mu)$ for $\Omega = \sigma(T)$ and some Borel measure μ, say $Tf = mf$ with $m \in L^\infty(\Omega, \mu)$. Note that the essential image of m equals $\sigma(T)$.

Assume first that T is injective, whence $\mu(m^{-1}(0)) = 0$. Using the principle value of the logarithm on $\mathbb{C} \setminus \{0\}$ which is measurable, define $(T(t))_{t \geq 0}$ by

$$T(t)f := e^{t \log m} f, \quad f \in L^2(\Omega, \mu), \ t \geq 0.$$

Each $T(t)$ is a bounded operator on $L^2(\Omega, \mu)$. Moreover, the family $(T(t))_{t \geq 0}$ is strongly continuous by Lebesgue's dominated convergence theorem. The semigroup law and the property $T(1) = T$ are clear, so T is embeddable.

Assume now $\dim \ker T = \infty$. Since T is normal, we have $\overline{\text{rg}\, T^*} = (\ker T)^\perp = (\ker T^*)^\perp = \overline{\text{rg}\, T}$. Therefore $\ker T$ and $\overline{\text{rg}\, T}$ reduce T and $H = \ker T \oplus \overline{\text{rg}\, T}$. The restriction $T|_{\ker T}$ is embeddable by Lemma 1.12. On the other hand, $T|_{\overline{\text{rg}\, T}}$ is a normal injective operator and hence embeddable by the first part of the proof. This shows that T is embeddable as well.

The remaining case is answered in the negative by Theorem 1.7.   □

By Theorem 1.14 and its proof we obtain the following well-known result.

Corollary 1.15. *Every unitary operator can be embedded into a unitary C_0-group with bounded generator.*

Remark 1.16. More generally, the proof of Theorem 1.14 shows that T can be embedded into a normal C_0-semigroup whenever T is injective. Moreover, if T is invertible, then 0 has positive distance to $\sigma(T)$ and T can be even embedded into a normal C_0-group with bounded generator employing the same definition for $t \in \mathbb{R}$.

Remark 1.17. Note that every operator similar to a normal operator is also embeddable. For example, Sz.-Nagy [236] showed that a bijective operator T on a Hilbert space satisfying $\sup_{j \in \mathbb{Z}} \|T^j\| < \infty$ is similar to a unitary operator. Hence such operators can be embedded into a C_0-group by Corollary 1.15. We refer also to van Casteren [45] for a characterisation of similarity to a self-adjoint operator. We finally refer to Benamara, Nikolski [29] for more on the similarity problem.

1.4   Isometries and co-isometries on Hilbert spaces

In this subsection we characterise the embedding property of isometries and co-isometries (i.e., operators with isometric adjoint) on Hilbert spaces. Note that the spectrum of a non-invertible isometry is the unit disc (see Conway [52, Exercise VII.6.7] for Banach spaces or Theorem II.5.5 for Hilbert spaces), and hence the spectral calculus method is not applicable.

The key is the Wold decomposition of isometries, see Theorem II.5.5. By this decomposition and Corollary 1.15, the question of embedding an isometry reduces to embedding the right shift on the space $l^2(\mathbb{N}, Y)$ for some Hilbert space Y. The following result shows how this can be achieved.

Proposition 1.18. *Let S be the right shift on $l^2(\mathbb{N}, Y)$ for an infinite-dimensional Hilbert space Y. Then S can be embedded into an isometric C_0-semigroup.*

Proof. Let S be the right shift on $l^2(\mathbb{N}, Y)$, i.e.,

$$S(x_1, x_2, x_3, \ldots) := (0, x_1, x_2, \ldots).$$

As before, Y is unitarily isomorphic to the space $L^2([0,1], Y)$, hence there is a unitary operator $J : l^2(\mathbb{N}, Y) \to l^2(\mathbb{N}, L^2([0,1], Y))$ such that JTJ^{-1} is again the right shift operator on $l^2(\mathbb{N}, L^2([0,1], Y))$. We now observe that $l^2(L^2([0,1], Y))$ can be identified with $L^2(\mathbb{R}_+, Y)$ by

$$(f_1, f_2, f_3, \ldots) \mapsto (s \mapsto f_n(s-n), \ s \in [n, n+1]).$$

By this identification the right shift operator on $l^2(\mathbb{N}, L^2([0,1], Y))$ corresponds to the operator

$$(Sf)(s) := \begin{cases} f(s-1), & s \geq 1, \\ 0, & s \in [0,1) \end{cases}$$

on $L^2(\mathbb{R}_+, Y)$ which can be embedded into the right shift semigroup on $L^2(\mathbb{R}_+, Y)$. Going back we see that our original operator S can be embedded into an isometric C_0-semigroup. $\qquad\square$

We can now characterise all isometries on Hilbert spaces having the embedding property.

Theorem 1.19. *An isometry V on a Hilbert space can be embedded into a C_0-semigroup if and only if V is unitary or $\mathrm{codim}(\mathrm{rg}\,V) = \infty$. In this case, one can embed V into an isometric C_0-semigroup.*

Proof. Let V be an isometry on a Hilbert space H. By the Wold decomposition we have the orthogonal decomposition $H = H_0 \oplus H_1$ into two invariant subspaces such that $V|_{H_0}$ is unitary and $V|_{H_1}$ is unitarily equivalent to the right shift on $l^2(\mathbb{N}, Y)$ for $Y := (\mathrm{rg}\,V)^\perp$.

By Corollary 1.15, we can embed $V|_{H_0}$ into a unitary C_0-group. If $Y = \{0\}$, the assertion follows. Now, if $\dim Y = \mathrm{codim}(\mathrm{rg}\,V) = \infty$, then we can embed $V|_{H_1}$ and therefore V by Proposition 1.18. Moreover, since the Wold decomposition is orthogonal and the semigroup from Proposition 1.18 is isometric, the constructed semigroup is isometric.

On the other hand, if $0 < \dim Y = \mathrm{codim}(\mathrm{rg}\,V) < \infty$, then V cannot be embedded into a C_0-semigroup by Theorem 1.7. $\qquad\square$

Since an operator on a Hilbert space is embeddable if and only if its adjoint is (see Engel, Nagel [78, Proposition I.5.14]), we also obtain the following characterisation for operators with isometric adjoint.

Corollary 1.20. *Let T be a co-isometry on a Hilbert space. Then T can be embedded into a C_0-semigroup if and only if T is injective or $\dim(\ker T) = \infty$.*

As an example we obtain that the left shift on $l^2(Y)$ for some Hilbert space Y has the embedding property if and only if $\dim(Y) = \infty$.

Open question 1.21. Does the necessary condition given in Theorem 1.7 suffice to embed partial isometries on Hilbert spaces?

Note that the answer is "yes" in each of the following cases:

(a) T is injective, i.e., T is an isometry (Theorem 1.19);

(b) T is surjective, i.e., T^* is an isometry (Corollary 1.20);

(c) $\ker T$ reduces T.

The argument in (c) is similar to the proof of Proposition 1.13 using Theorem 1.19.

1.5 Abstract examples: a residuality result

We now show that the embedding property is natural for operators, at least on Hilbert spaces. More precisely, we show that a "typical" (in the sense of Baire categories) contraction on a separable Hilbert space is embeddable.

The key to our result is the following theorem showing that a "typical" contraction as well as a "typical" isometry is unitary which is also of independent interest. This is a consequence of Theorem IV.3.9 and Remark IV.3.2, but we give here a direct proof following [66].

Theorem 1.22. *Let H be a separable infinite-dimensional Hilbert space. Then the set $\mathcal{U}$ of all unitary operators is residual in the set $\mathcal{C}$ of all contractions for the weak operator topology. In addition, $\mathcal{U}$ is residual in the set $\mathcal{I}$ of all isometries for the strong operator topology as well.*

Recall that the space of all contractions on H is a complete metric space for the weak operator topology, see Subsection II.5.3.

Proof. The proof is divided into two parts. We first show that a "typical" isometry is unitary for the strong operator topology and then consider contractions.

Part 1: isometries. We prove residuality of $\mathcal{U}$ in the set of all isometries $\mathcal{I}$ on H endowed with the strong operator topology.

Fix a dense subset $\{x_j\}_{j=1}^{\infty} \subset H \setminus \{0\}$ and let T be a non-invertible isometry. Then $\operatorname{rg} T$ is closed and different from H. Therefore there exists x_j with $\operatorname{dist}(x_j, \operatorname{rg} T) > 0$, hence

$$\mathcal{I} \setminus \mathcal{U} = \bigcup_{k,j=1}^{\infty} M_{j,k} \quad \text{with} \quad M_{j,k} := \left\{ T : \operatorname{dist}(x_j, \operatorname{rg} T) > \frac{1}{k} \right\}.$$

We now prove that every set $M_{j,k}$ is nowhere dense in $\mathcal{I}$. By Proposition II.5.7, unitary operators are dense in $\mathcal{I}$ and therefore it suffices to show that

$$\mathcal{U} \cap \overline{M_{j,k}} = \emptyset \quad \forall j, k. \tag{V.3}$$

Assume the contrary, i.e., that there exists a sequence $\{T_n\}_{n=1}^{\infty} \subset M_{j,k}$ for some j, k and a unitary operator U with $\lim_{n\to\infty} T_n = U$ strongly. In particular, $\lim_{n\to\infty} T_n y = U y = x_j$ for $y := U^{-1} x_j$. This however implies the contradiction $\lim_{n\to\infty} \text{dist}(x_j, \text{rg}\, T_n) = 0$. So (V.3) is proved, every set $M_{j,k}$ is nowhere dense, and $\mathcal{U}$ is residual in $\mathcal{I}$.

Part 2: contractions. We first prove that the set $\mathcal{I} \setminus \mathcal{U}$ of non-invertible isometries is of first category in $\mathcal{C}$. As in Part 1, $\mathcal{I} \setminus \mathcal{U}$ is given as

$$\mathcal{I} \setminus \mathcal{U} = \bigcup_{k,j=1}^{\infty} M_{j,k} \quad \text{with} \quad M_{j,k} := \left\{ T \text{ isometric} : \text{dist}(x_j, \text{rg}\, T) > \frac{1}{k} \right\}.$$

Since unitary operators are dense in $\mathcal{C}$, see Takesaki [239, p. 99] or Peller [211], it is enough to show that

$$\mathcal{U} \cap \overline{M_{j,k}} = \emptyset \quad \forall j, k.$$

Assume that for some j, k there exists a sequence $\{T_n\}_{n=1}^{\infty} \subset M_{j,k}$ converging weakly to a unitary operator U. Then, by Lemma II.5.11, T_n converges to U strongly. As in Part 1, this implies $\lim_{n\to\infty} T_n y = U y = x_j$ for $y := U^{-1} x_j$. Hence $\lim_{n\to\infty} \text{dist}(x_j, \text{rg}\, T_n) = 0$ contradicting $\{T_n\}_{n=1}^{\infty} \subset M_{j,k}$, so every $M_{j,k}$ is nowhere dense and $\mathcal{I} \setminus \mathcal{U}$ is of first category.

We now show that the set of non-isometric operators is of first category in $\mathcal{C}$ as well. Let T be a non-isometric contraction. Then there exists x_j such that $\|T x_j\| < \|x_j\|$, hence

$$\mathcal{C} \setminus \mathcal{I} = \bigcup_{k,j=1}^{\infty} N_{j,k} \quad \text{with} \quad N_{j,k} := \left\{ T : \frac{\|T x_j\|}{\|x_j\|} < 1 - \frac{1}{k} \right\}.$$

It remains to show that every $N_{j,k}$ is nowhere dense in $\mathcal{C}$. By density of $\mathcal{U}$ in $\mathcal{C}$ it again suffices to show that

$$\mathcal{U} \cap \overline{N_{j,k}} = \emptyset \quad \forall j, k.$$

Assume that for some j, k there exists a sequence $\{T_n\}_{n=1}^{\infty} \subset N_{j,k}$ converging weakly to a unitary operator U. Then T_n converges to U strongly by Lemma II.5.11. This implies that $\lim_{n\to\infty} \|T_n x_j\| = \|U x_j\| = \|x_j\|$ contradicting $\frac{\|T_n x_j\|}{\|x_j\|} < 1 - \frac{1}{k}$ for every $n \in \mathbb{N}$. $\qquad\square$

Remark 1.23. It is an interesting question for which W*-algebras the corresponding result holds, i.e., the unitary elements are residual in the unit ball. Note that l^{∞} does not have this property since the unitary elements (i.e., unimodular sequences) are not dense in the unit ball for the weak* topology.

Combining the above theorem with Corollary 1.15, we obtain that the embedding property is "typical" for such operators.

Theorem 1.24. *Let H be a separable infinite-dimensional Hilbert space. Then the set of all embeddable contractions on H is residual in the set of all contractions for the weak operator topology. In addition, the set of all embeddable isometries is residual in the set of all isometries for the strong operator topology.*

Remark 1.25. This result is an operator-theoretical counterpart to a recent result of de la Rue and de Sam Lazaro [59] in ergodic theory stating that a "typical" measure preserving transformation can be embedded into a measure preserving flow. We refer to Stepin, Eremenko [235] for further results and references.

In particular, a "typical" contraction or isometry on a separable infinite-dimensional Hilbert space has roots of all orders, which is an operator-theoretical analogue of a result of King [149] in ergodic theory.

To finish this section we make the embedding problem more difficult, but more relevant to ergodic and measure theory.

Open question 1.26. Let T be a positive operator on a Banach lattice. When is T embeddable into a positive C_0-semigroup?

2   Cogenerators

In this section we study connections between a C_0-semigroup $(T(t))_{t \geq 0}$ and the discrete system $\{V^n\}_{n=0}^{\infty}$ obtained from the cogenerator $V = -(I + A)R(1, A)$ of $T(\cdot)$. For basic properties of the cogenerator see Subsection I.3.1. We only recall that V is a bounded operator determining the semigroup uniquely. It is obtained from the resolvent of the generator and hence does not involve any explicit knowledge of the semigroup.

We already encountered in Subsection I.3.1 some symmetries between $(T(t))_{t \geq 0}$ and V on Hilbert spaces for properties such as contractivity and unitarity. On Banach spaces, however, not so many results in this direction are known. The aim of this section is to describe the general situation and state the known results and some open problems. We mainly follow Eisner, Zwart [74].

2.1   (Power) boundedness

We first discuss the connection between boundedness of $T(\cdot)$ and power boundedness of V.

We start with a Hille–Yosida type characterisation for contractive and bounded semigroups on Banach spaces using the behaviour of the resolvent of V near the point 1.

Theorem 2.1. *For an operator V on a Banach space X, the following assertions are equivalent.*

(i) *V is the cogenerator of a contraction C_0-semigroup on X.*

(ii) *$V - I$ is injective and has dense range; $(1, \infty) \in \rho(V)$ and*

$$\|(I - V)R(\mu, V)\| \leq \frac{2}{\mu + 1} \quad \text{for all } \mu > 1. \tag{V.4}$$

(iii) *$V - I$ is injective and has dense range; there exists $\mu_0 > 1$ such that $(1, \mu_0) \in \rho(V)$ and*

$$\|(I - V)R(\mu, V)\| \leq \frac{2}{\mu + 1} \quad \text{for all } \mu \in (1, \mu_0). \tag{V.5}$$

Proof. We first note that injectivity and dense range of the operator $V - I$ is necessary for every cogenerator V. Assume now $V - I$ to be injective and to have dense range.

Define $A := (V + I)(V - I)^{-1}$ which is densely defined. By the Hille–Yosida theorem A generates a contraction semigroup if and only if $(\lambda_0, \infty) \subset \rho(A)$ for some $\lambda_0 \geq 0$ and $\|R(\lambda, A)\| \leq \frac{1}{\lambda}$ holds for all $\lambda > \lambda_0$. Note that $\lambda > \lambda_0 \geq 0$ holds if and only if $1 < \mu \leq \mu_0$ for $\mu := \frac{\lambda+1}{\lambda-1}$ and $\mu_0 := \frac{\lambda_0+1}{\lambda_0-1}$. Moreover, by Proposition I.3.3, we have for $1 < \mu \in \rho(V)$ that $0 < \lambda := \frac{\mu+1}{\mu-1} \in \rho(A)$ and

$$\lambda R(\lambda, A) = \frac{\lambda}{\lambda - 1}(I - V)R(\mu, V) = \frac{\mu + 1}{2}(I - V)R(\mu, V). \tag{V.6}$$

This proves the equivalence of (i) and (iii). Using the same arguments one shows (i)$\Leftrightarrow$(ii). $\qquad\square$

Analogously, one can treat boundedness.

Theorem 2.2. *For $V \in \mathcal{L}(X)$ and $M \geq 1$, the following assertions are equivalent.*

(i) *V is the cogenerator of a C_0-semigroup $(T(t))_{t \geq 0}$ on X satisfying $\|T(t)\| \leq M$ for all $t \geq 0$.*

(ii) *$V - I$ is injective and has dense range; $(1, \infty) \in \rho(V)$ and*

$$\| [(I - V)R(\mu, V)]^n \| \leq \frac{2^n M}{(\mu + 1)^n} \quad \text{for all } \mu > 1, \; n \in \mathbb{N}. \tag{V.7}$$

(iii) *$V - I$ is injective and has dense range; there exists $\mu_0 > 1$ such that $(1, \mu_0) \in \rho(V)$ and*

$$\| [(I - V)R(\mu, V)]^n \| \leq \frac{2^n M}{(\mu + 1)^n} \quad \text{for all } \mu \in (1, \mu_0), \; n \in \mathbb{N}. \tag{V.8}$$

In all these characterisations the resolvent $R(\lambda, V)$ of the cogenerator had to be used. In fact, the boundedness of $T(\cdot)$ is not equivalent to the power boundedness of V. In order to prove some results in this direction we first need a technical lemma on Laguerre polynomials whose proof (in a more general case) can be found in [74].

Lemma 2.3. *Let $L_n^1(t)$ denote the first generalised Laguerre polynomial, i.e.,*

$$L_n^1(t) = \sum_{m=0}^{n} \frac{(-1)^m}{m!} \binom{n+1}{n-m} t^m. \tag{V.9}$$

Then the inequality

$$C_1 \sqrt{n} \leq \int_0^\infty |L_n^1(2t)| e^{-t} dt \leq C_2 \sqrt{n} \tag{V.10}$$

holds for some constants C_1, C_2 and all $n \in \mathbb{N}$.

We are now ready to prove the result of Brenner, Thomée [41] giving the upper bound for the growth of the powers of the cogenerator of a bounded C_0-semigroup. We follow the proof in [74].

Theorem 2.4. *Let $T(\cdot)$ be a bounded C_0-semigroup on a Banach space with cogenerator V. Then $\|V^n\| \leq C\sqrt{n}$ some C and every $n \in \mathbb{N}$.*

Furthermore, this estimate cannot be improved, i.e., there exists a Banach space and a bounded C_0-semigroup such that $\|V^n\| \geq C\sqrt{n}$ for some $C > 0$ and every $n \in \mathbb{N}$.

Proof. Using Lemma I.3.6 and boundedness of the semigroup we observe that

$$\|V^n\| \leq 1 + 2M \int_0^\infty |L_{n-1}^1(2t)| e^{-t} dt \tag{V.11}$$

for $M := \sup_{t \geq 0} \|T(t)\|$ and all $n \in \mathbb{N}$. Lemma 2.3 immediately implies $\|V^n\| \leq C\sqrt{n}$ for some C and every $n \in \mathbb{N}$.

We now show that this estimate is sharp. Consider $X := C_0([0, \infty))$ and the left shift semigroup $T(\cdot)$ on X given by $(T(t)f)(s) = f(t + s)$. Note that $T(\cdot)$ is contractive and the powers of its cogenerator V are given by the formula

$$(V^n f)(s) = f(s) - 2 \int_0^\infty L_{n-1}^1(2t) e^{-t} f(t + s) dt$$

by Lemma I.3.6. Define now $h(s) := \mathrm{sign}(L_{n-1}^1(2s))$. We obtain formally that

$$(V^n h)(s) = \mathrm{sign}(L_{n-1}^1(2s)) - 2 \int_0^\infty |L_{n-1}^1(2t)| e^{-t} dt,$$

and hence

$$|(V^n h)(s)| \geq C\sqrt{n} \text{ for every } s \geq 0$$

by Lemma 2.3. It remains to approximate h by continuous functions. Thus, for every $\varepsilon > 0$ there exists a function $f_\varepsilon \in C_0([0, \infty))$ with $\|f\| = 1$ such that the above estimate holds within an error of at most ε, which finishes the proof. $\square$

Remark 2.5. By the standard renorming procedure (see Lemma III.1.4), one derives from the above an example of a contractive C_0-semigroup on a Banach space such that the powers of its cogenerator grow as $\sqrt{n}$. This shows that the Foiaş–Sz.-Nagy theorem (see Subsection I.3.1) fails in Banach spaces. For more examples see Subsection 2.3 below.

On Hilbert spaces the following question is still open.

Open question 2.6. Let $T(\cdot)$ be a C_0-semigroup on a Hilbert space with cogenerator V. Is boundedness of $T(\cdot)$ equivalent to power boundedness of V?

Guo, Zwart [113] gave a partial answer to this question under additional assumptions.

Theorem 2.7 (Guo, Zwart). *Let $T(\cdot)$ be a C_0-semigroup on a Hilbert space with generator A and cogenerator V. Then the following assertions hold.*

(a) *If $T(\cdot)$ is bounded and A^{-1} generates a bounded C_0-semigroup as well, then V is power bounded.*

(b) *If $T(\cdot)$ is analytic, then its boundedness is equivalent to power boundedness of V.*

We refer to Guo, Zwart [113] for the proofs and further results.

It is not clear whether assertion (a) in the above theorem holds for semigroups on Banach spaces as well.

2.2 Characterisation via cogenerators of the rescaled semigroups

In this subsection we characterise boundedness of a C_0-semigroup using the cogenerators of the rescaled semigroups.

We begin with the following observation. If A generates a contractive or bounded C_0-semigroup, then all operators τA, $\tau > 0$, do so as well. However, as we will see in Subsection 2.3, it is not always true that the operators

$$V_\tau := (\tau A + I)(\tau A - I)^{-1}, \quad \tau > 0, \tag{V.12}$$

remain contractive when V is, so one needs another condition to obtain equivalence.

Proposition 2.8. *For a densely defined operator A on a Banach space X, the following assertions are equivalent.*

(i) *A generates a contraction C_0-semigroup on X.*

(ii) $(0, \infty) \subset \rho(A)$ and the operators V_τ satisfy

$$\|V_\tau - I\| \leq 2 \quad \text{for all } \tau > 0.$$

(iii) There exists $\tau_0 > 0$ such that $(\frac{1}{\tau_0}, \infty) \subset \rho(A)$ and the operators V_τ satisfy

$$\|V_\tau - I\| \leq 2 \quad \text{for all } 0 < \tau < \tau_0.$$

Proof. By the formula

$$V_\tau = (A + tI)(A - tI)^{-1} = I - 2tR(t, A)$$

for $t := \frac{1}{\tau}$ we immediately obtain that

$$tR(t, A) = \frac{I - V_\tau}{2}. \tag{V.13}$$

Then the proposition follows from the Hille–Yosida theorem. $\square$

Remark 2.9. The following nice property holds for cogenerators of C_0-semigroups on Hilbert spaces: Contractivity of V automatically implies contractivity of every V_τ, $\tau > 0$. This follows from

$$\|V_\tau x\|^2 - \|x\|^2 = \|(A + tI)(A - tI)^{-1} x\|^2 - \|x\|^2$$
$$= \langle (A + tI)y, (A + tI)y \rangle - \langle (A - tI)y, (A - tI)y \rangle = 4t\mathrm{Re}\, \langle Ay, y \rangle$$

for $y := (A - tI)^{-1}x$. As we will see in Subsection 2.3, this property fails on Banach spaces.

A result analogous to Proposition 2.8 holds for generators of bounded C_0-semigroups as well.

Proposition 2.10. *For a densely defined operator A on a Banach space, the following assertions are equivalent.*

(i) *A generates a C_0-semigroup $(T(t))_{t \geq 0}$ satisfying $\|T(t)\| \leq M$ for every $t \geq 0$.*

(ii) *$(0, \infty) \subset \rho(A)$ and the operators V_τ satisfy*

$$\left\| \left[\frac{V_\tau - I}{2} \right]^n \right\| \leq M \quad \text{for all } \tau > 0 \text{ and } n \in \mathbb{N}.$$

(iii) *There exists τ_0 such that $(\frac{1}{\tau_0}, \infty) \subset \rho(A)$ and the operators V_τ satisfy*

$$\left\| \left[\frac{V_\tau - I}{2} \right]^n \right\| \leq M \quad \text{for all } 0 < \tau < \tau_0 \text{ and } n \in \mathbb{N}.$$

The proof follows from formula (V.13) and the Hille–Yosida theorem for bounded semigroups.

Propositions 2.8 and 2.10 imply the following sufficient condition being analogous to the one of Foiaş and Sz.-Nagy in Subsection I.3.1.

Theorem 2.11. *For a densely defined operator A on a Banach space X, the following assertions hold.*

(a) *If there exists $\tau_0 > 0$ such that $(\frac{1}{\tau_0}, \infty) \subset \rho(A)$ and the operators V_τ are contractive for every $\tau \in (0, \tau_0)$, then A generates a contractive C_0-semigroup.*

(b) *If there exists $\tau_0 > 0$ such that $(\frac{1}{\tau_0}, \infty) \subset \rho(A)$ and the operators V_τ satisfy $\|V_\tau^n\| \leq M$ for all $\tau \in (0, \tau_0)$ and $n \in \mathbb{N}$, then A generates a C_0-semigroup $(T(t))_{t \geq 0}$ with $\|T(t)\| \leq M$ for all $t \geq 0$.*

Proof. Assertion (a) follows immediately from Proposition 2.8. To prove (b) assume that $\|V_\tau^n\| \leq M$. Then we have

$$\left\| \left[\frac{V_\tau - I}{2} \right]^n \right\| \leq \frac{1}{2^n} \sum_{j=0}^{n} \binom{n}{j} \|V_\tau^j\| \leq \frac{M \cdot 2^n}{2^n} = M$$

and (b) follows from Proposition 2.10. □

Remark 2.12. In Proposition 2.8, Proposition 2.10 and Theorem 2.11 it suffices to consider $\{V_{\tau_n}\}_{n=1}^{\infty}$ for a sequence $\tau_n > 0$ converging to zero. This again follows from the fact that in the Hille–Yosida theorem it suffices to check the resolvent condition only for a sequence $\{\lambda_n\}_{n=1}^{\infty} \subset (0, \infty)$ converging to infinity.

We finally observe the following. If V is contractive or power bounded, then so is the operator $-V$. Note that $-V$ is the cogenerator of the semigroup generated by A^{-1} if A^{-1} generates a C_0-semigroup. However, contractivity or boundedness of $(e^{tA})_{t \geq 0}$ does not imply the same property of $(e^{tA^{-1}})_{t \geq 0}$ (see Zwart [265] and also Subsection 2.3 for elementary examples). We refer to Zwart [265, 266], Gomilko, Zwart [107], Gomilko, Zwart, Tomilov [108], de Laubenfels [60] for further information on this aspect.

Remark 2.13. Assume that $(0, \infty) \subset \rho(A)$ and A^{-1} exists as a densely defined operator. Then we have

$$V_{\tau, A^{-1}} = (\tau A^{-1} + I)(\tau A^{-1} - I)^{-1} = (\tau I + A)(\tau I - A)^{-1} = -V_{\frac{1}{\tau}, A}. \qquad (V.14)$$

So we see that contractivity (power boundedness) of V_τ for *all* $\tau > 0$ or even for sequences $\tau_{n,1} \to 0$ and $\tau_{n,2} \to \infty$ implies that A and A^{-1} both generate a contractive (bounded) C_0-semigroup.

Conversely, Gomilko [105] and Guo, Zwart [113] showed for Hilbert spaces that if A and A^{-1} both generate bounded semigroups, then the cogenerator V is power bounded (see Theorem 2.7) and hence so are all operators V_τ by the rescaling argument.

2.3 Examples

In this subsection we discuss examples describing various situations on Banach spaces and show in particular that the Foiaş–Sz.-Nagy theorem (see Subsection I.3.1) fails even on 2-dimensional Banach spaces.

In [108] Gomilko, Zwart and Tomilov show that for every $p \in [1, \infty)$, $p \neq 2$, there exists a contractive operator V on l^p such that $(V - I)^{-1}$ exists as a densely defined operator, but V is not the cogenerator of a C_0-semigroup. The idea of their construction is the following. Take the bounded operator $A := S_l - I$, where S_l denotes the left shift given by $S_l(x_1, x_2, x_3, \ldots) = (x_2, x_3, \ldots)$. The cogenerator V of the semigroup generated by A is given by $V = S_l R(2, S_l)$ and hence is contractive (use contractivity of S_l and the Neumann series for the resolvent). Further, one shows that A^{-1} does not generate a C_0-semigroup which is the hard part. As a consequence one obtains that the contraction $-V$ is not the cogenerator of a C_0-semigroup.

Komatsu [152, pp. 343–344] showed that the operator $A := S_r - I$ for the right shift S_r given by $S_r(x_1, x_2, x_3, \ldots) = (0, x_1, x_2, \ldots)$ on c_0 satisfies the same properties, i.e., A^{-1} does not generate a C_0-semigroup. Since the cogenerator V corresponding to A is contractive as well, we have a contraction on c_0 which is not the cogenerator of a C_0-semigroup.

The following example shows that even for $X = \mathbb{C}^2$ the semigroup cogenerated by a contraction need not be contractive. In particular, this example and Example 2.16 show that no implication in the Foiaş–Sz.-Nagy theorem holds on two-dimensional Banach spaces.

Example 2.14. Take $X = \mathbb{C}^2$ endowed with $\| \cdot \|_p$, $p \neq 2$, and $A := \begin{pmatrix} -1 & \beta \\ 0 & -2 \end{pmatrix}$ for $\beta > 0$. The semigroup generated by A is

$$T(t) = \begin{pmatrix} e^{-t} & \beta(e^{-t} - e^{-2t}) \\ 0 & e^{-2t} \end{pmatrix}, \quad t \geq 0.$$

We first show that $(T(t))_{t \geq 0}$ is not contractive for appropriate β. Consider first $p = \infty$ and $\beta > 1$. We have $\|T(t)\| = (1 + \beta)e^{-t} - \beta e^{-2t} =: f(t)$. Since $f(0) = 1$ and $f'(0) = \beta - 1 > 0$, the semigroup is not contractive.

Let now $2 < p < \infty$ and define $\beta := (3^p - 1)^{\frac{1}{p}}$. Then

$$\left\| T(t) \begin{pmatrix} x \\ 1 \end{pmatrix} \right\|_p^p = (e^{-t}x + \beta(e^{-t} - e^{-2t}))^p + e^{-2pt} =: f_x(t) \quad \text{for } x > 0.$$

We have $f_x(0) = x^p + 1 = \|(x, 1)\|_p^p$. Further, $f_x'(0) = px^{p-1}(\beta - x) - 2p$, so the semigroup is not contractive if $x^{p-1}(\beta - x) > 2$ for some $x > 0$. This is the case for $x := \frac{\beta}{2}$. Indeed, $x^{p-1}(\beta - x) = \left(\frac{\beta}{2}\right)^p = \frac{3^p - 1}{2^p} > 2$.

We now show that the cogenerator V is contractive for $\beta \leq 3$ if $p = \infty$ and $\beta := (3^p - 1)^{\frac{1}{p}}$ if $p \in (2, \infty)$. The cogenerator is given by

$$V = (I + A)(A - I)^{-1} = \begin{pmatrix} 0 & \beta \\ 0 & -1 \end{pmatrix} \begin{pmatrix} -\frac{1}{2} & -\frac{\beta}{6} \\ 0 & -\frac{1}{3} \end{pmatrix} = \begin{pmatrix} 0 & -\frac{\beta}{3} \\ 0 & \frac{1}{3} \end{pmatrix}.$$

So for $p = \infty$ we have $\|V\| = \max\{\frac{1}{3}, \frac{\beta}{3}\} \leq 1$ for $\beta \leq 3$. For $p \in (2, \infty)$ we have
$\|V\|^p = \|(-\frac{\beta}{3}, \frac{1}{3})\|_p^p = \frac{(\beta^p + 1)}{3^p} \leq 1$ if and only if $\beta \leq (3^p - 1)^{\frac{1}{p}}$.

We see that for $p \in (2, \infty]$ there exists a contraction such that the cogenerated semigroup is not contractive. The analogous assertion for $p \in [1, 2)$ follows by duality.

Remark 2.15. From Theorem 2.11, Remark 2.12 and the above example we see that there exist contractions V (even on $\mathbb{C}^2$ with l^p-norm, $p \neq 2$) such that the operators V_τ are not contractive for every τ in a small interval $(0, \tau_0)$.

The following example gives a class of contractive semigroups on $(\mathbb{C}^2, \|\cdot\|_\infty)$ with non-contractive cogenerators. In particular, this provides a two-dimensional counterexample to the converse implication in the Foiaş–Sz.-Nagy theorem.

Example 2.16. Every operator A generating a contractive C_0-semigroup such that A^{-1} generates a C_0-semigroup which is not contractive leads to an example of a contractive semigroup with non-contractive cogenerator. Indeed, by the previous remark, there exists $\tau > 0$ such that V_τ is not contractive. Therefore, the operator τA generates a contractive semigroup with non-contractive cogenerator.

For a concrete example consider $X := \mathbb{C}^2$ endowed with $\|\cdot\|_\infty$ and A as in Example 2.14. Then $(e^{tA})_{t \geq 0}$ is not contractive for $\beta > 1$. We show that the semigroup generated by A^{-1} is contractive if and only if $\beta \leq 2$.

Indeed, we have $A^{-1} = \begin{pmatrix} -1 & -\frac{\beta}{2} \\ 0 & -\frac{1}{2} \end{pmatrix}$ and

$$e^{tA^{-1}} = \begin{pmatrix} e^{-t} & \beta(e^{-t} - e^{-\frac{t}{2}}) \\ 0 & e^{-\frac{t}{2}} \end{pmatrix}, \quad t \geq 0.$$

Therefore $\|e^{tA^{-1}}\|_\infty = \sup\{e^{-t} + \beta(e^{-\frac{t}{2}} - e^{-t}), e^{-\frac{t}{2}}\}$. Hence $e^{tA^{-1}}$ is contractive if and only if $g(t) := e^{-t} + \beta(e^{-\frac{t}{2}} - e^{-t}) \leq 1$ for every $t > 0$. We have $g(0) = 1$ and $g'(t) = -e^{-t} + \beta(e^{-t} - \frac{1}{2}e^{-\frac{t}{2}}) = e^{-t}[\beta(1 - \frac{1}{2}e^{\frac{t}{2}}) - 1]$. Since the function $t \to 1 - \frac{1}{2}e^{\frac{t}{2}}$ is monotonically decreasing, we obtain that $g'(t) \leq 0$ for every $t \geq 0$ is equivalent to $g'(0) \leq 0$, i.e., $\beta \leq 2$.

So we see that for $1 < \beta \leq 2$ the semigroup generated by A^{-1} is contractive while the semigroup generated by A is not contractive. The rescaling procedure described above leads to a contractive semigroup (generated by τA for some τ) with non-contractive cogenerator.

Zwart [265] gives another example of an operator A generating a contractive C_0-semigroup such that the semigroup generated by A^{-1} is not contractive and not even bounded. He takes a nilpotent semigroup on $X = C_0[0, 1]$ such that the semigroup generated by A^{-1} grows like $t^{1/4}$. By the rescaling procedure we again obtain a contractive semigroup with non-contractive cogenerator.

Remark 2.17. The above example for $2 < \beta \leq 3$ yields a contractive cogenerator V such that the semigroups generated by both operators A and A^{-1} are not contractive. So V is a contraction on $(\mathbb{C}^2, \|\cdot\|_\infty)$ such that operators V_τ are not contractive for every $\tau \in (0, \tau_1) \cup (\tau_2, \infty)$, $0 < \tau_1 < 1 < \tau_2$, by Remark 2.12.

2.4 Polynomial boundedness

In this section, again following Eisner, Zwart [74], we show how polynomial boundedness of a C_0-semigroup can be characterised by its cogenerator.

We show first that polynomial boundedness of a C_0-semigroup on a Banach space implies polynomial boundedness of its cogenerator. This generalises results of Hersh and Kato [126] and Brenner and Thomée [41] on bounded semigroups. For the proof, which is analogous to the one of Theorem 2.4, see [74].

Theorem 2.18. *Let $T(\cdot)$ be a C_0-semigroup on a Banach space with cogenerator V. If $\|T(t)\| \le M(1+t^k)$ for some M and $k \in \mathbb{N} \cup \{0\}$ and every $t \ge 0$, then $\|V^n\| \le Cn^{k+\frac{1}{2}}$ some C and every $n \in \mathbb{N}$.*

Furthermore, this estimate cannot be improved, i.e., for every $k \in \mathbb{N} \cup \{0\}$ there exists a Banach space and a C_0-semigroup satisfying $\|T(t)\| = O(t^k)$ such that $\|V^n\| \ge Cn^{k+\frac{1}{2}}$ for some $C > 0$ and every $n \in \mathbb{N}$.

Our next result shows that for analytic semigroups the converse implication in Theorem 2.18 holds, i.e., polynomial boundedness of the cogenerator implies polynomial boundedness of the semigroup.

Theorem 2.19. *Let $T(\cdot)$ be an analytic C_0-semigroup on a Banach space with cogenerator V. If $\|V^n\| \le Cn^k$ for some $C, k \ge 0$ and every $n \in \mathbb{N}$, then $\|T(t)\| \le M(1+t^{2k+1})$ for some M and every $t \ge 0$.*

Proof. Assume $\|V^n\| \le Cn^k$ for some $C, k \ge 0$ and every $n \in \mathbb{N}$. Then $r(V) \le 1$ and therefore $\lambda \in \rho(A)$ for $\mathrm{Re}\,\lambda > 0$ by Proposition I.3.3, where A denotes the generator of $T(\cdot)$. Our aim is to show that there exist $a_0, M > 0$ such that

$$\|R(\lambda, A)\| \le \frac{M}{(\mathrm{Re}\,\lambda)^{k+1}} \text{ for all } \lambda \text{ with } 0 < \mathrm{Re}\,\lambda < a_0, \qquad (V.15)$$

$$\|R(\lambda, A)\| \le M \text{ for all } \lambda \text{ with } \mathrm{Re}\,\lambda \ge a_0. \qquad (V.16)$$

By Theorem III.1.20 this implies growth of $\|T(t)\|$ at most as t^{2k+1} for analytic semigroups.

Take some $a_0 > \max\{0, \omega_0(T)\}$. Then (V.16) automatically holds and we only have to show (V.15).

Since $T(\cdot)$ is analytic, $R(\lambda, A)$ is uniformly bounded on $\{\lambda : |\mathrm{Im}\,\lambda| > b_0, \ 0 < \mathrm{Re}\,\lambda < a_0\}$ for some $b_0 \ge 0$. Moreover, $R(\lambda, A)$ is also uniformly bounded on $\{\lambda : |\mathrm{Im}\,\lambda| \le b_0, \ \frac{1}{3} \le \mathrm{Re}\,\lambda < a_0\}$ as well. Take now λ with $0 < \mathrm{Re}\,\lambda < \frac{1}{3}$ and $-b_0 < \mathrm{Im}\,\lambda < b_0$.

By Proposition I.3.3, where A denotes the generator of $T(\cdot)$ we have

$$\|R(\lambda, A)\| \le \frac{1 + \|V\|}{|\lambda - 1|} \left\| R\left(\frac{\lambda + 1}{\lambda - 1}, V\right) \right\|. \qquad (V.17)$$

By Theorem II.1.17, growth of $\|V^n\|$ like n^k implies

$$\|R(\mu, V)\| \le \frac{\tilde{C}}{(|\mu| - 1)^{k+1}} \text{ for all } \mu \text{ with } 1 < |\mu| \le 2 \qquad (V.18)$$

for some constant $\tilde{C}$. For $\mu := \frac{\lambda+1}{\lambda-1}$ and $0 < \operatorname{Re}\lambda < \frac{1}{3}$ we have

$$|\mu| - 1 = \frac{|\lambda+1| - |\lambda-1|}{|\lambda-1|} = \frac{4\operatorname{Re}\lambda}{|\lambda-1|(|\lambda+1| + |\lambda-1|)} < 1.$$

Then we use (V.18) to obtain

$$\left\| R\left(\frac{\lambda+1}{\lambda-1}, V \right) \right\| \leq \frac{\tilde{C}|\lambda-1|^{k+1}(|\lambda+1| + |\lambda-1|)^{k+1}}{4^{k+1}(\operatorname{Re}\lambda)^{k+1}} \leq \frac{C_1}{(\operatorname{Re}\lambda)^{k+1}}$$

for $C_1 := \tilde{C}(b_0^2 + \frac{5}{4})^{k+1}$. So, by (V.17),

$$\|R(\lambda, A)\| \leq \frac{C_1(1 + \|V\|)}{|\lambda-1|(\operatorname{Re}\lambda)^{k+1}} \leq \frac{C_1(1 + \|V\|)}{2(\operatorname{Re}\lambda)^{k+1}}$$

which proves (V.15). $\qquad\qquad\qquad\qquad\qquad\qquad\qquad\qquad\qquad\square$

2.5 Strong stability

We now discuss the relation between strong stability of a C_0-semigroup and its cogenerator restricting our attention to Hilbert spaces.

The following classical theorem is based on the dilation theory developed by Foiaş and Sz.-Nagy, see their monograph [238].

Theorem 2.20 (Foiaş, Sz.-Nagy [238, Prop. III.9.1]). *Let $T(\cdot)$ be a contraction semigroup on a Hilbert space H with cogenerator V. Then*

$$\lim_{t\to\infty} \|T(t)x\| = \lim_{n\to\infty} \|V^n x\|$$

holds for every $x \in H$. In particular, $T(\cdot)$ is strongly stable if and only if its cogenerator V is strongly stable.

Guo and Zwart obtained a (partial) generalisation of Theorem 2.20 to bounded C_0-semigroups.

Theorem 2.21 (Guo, Zwart, [113]). *Let $T(\cdot)$ be a bounded semigroup on a Hilbert space H with power bounded cogenerator V. If $T(\cdot)$ is strongly stable, then so is V.*

Note that the assumption of power boundedness of V in the above theorem is satisfied if for example A^{-1} exists and generates a bounded C_0-semigroup as well, see Theorem 2.7.

No such result seems to be known on Banach spaces.

2.6   Weak and almost weak stability

To conclude we look at the relation between weak and almost weak stability of a C_0-semigroup and its cogenerator.

We first show that for a large class of semigroups almost weak stability is preserved by the cogenerator.

Proposition 2.22. *Let $T(\cdot)$ be a relatively weakly compact C_0-semigroup on a Banach space such that its cogenerator V has relatively weakly compact orbits. Then $T(\cdot)$ is almost weakly stable if and only if V is.*

In particular, a bounded C_0-semigroup on a reflexive Banach space with power bounded cogenerator is almost weakly stable if and only if its cogenerator is.

Proof. Denote the generator and the cogenerator of $T(\cdot)$ by A and V, respectively. By Proposition I.3.3 we have

$$P_\sigma(A) \cap i\mathbb{R} = \emptyset \quad \text{if and only if} \quad P_\sigma(V) \cap \Gamma = \emptyset.$$

Theorems II.4.1 and III.5.1 conclude the argument.

The last assertion follows from Example I.1.7 (a).   □

Open question 2.23. Let $T(\cdot)$ be a contractive C_0-semigroup on a Hilbert space. Is weak stability of $T(\cdot)$ equivalent to weak stability of its cogenerator?

Remark 2.24. By Theorems II.3.13 and III.4.10, one can decompose every contraction (semigroup) on a Hilbert space into a unitary and a completely non-unitary part, where the completely non-unitary part is always weakly stable. Moreover, a semigroup is contractive or unitary if and only if its cogenerator is (see Subsection I.3.1).

Therefore the question above restricts to the case of unitary groups and, by the considerations in Section IV.1, can be reformulated as follows. Assume μ to be a probability measure on $\mathbb{R}$. Denote by ν the image of μ under the mapping

$$z \mapsto \frac{iz+1}{iz-1}, \quad \mathbb{R} \to \Gamma.$$

Does the following equivalence hold:

$$\mu \text{ is Rajchman on } \mathbb{R} \iff \nu \text{ is Rajchman on } \Gamma?$$

Bibliography

[1] O. N. Ageev, *On the genericity of some non-asymptotic dynamical proper-ties*, Uspekhi Mat. Nauk **58** (2003), 177–178; translation in Russian Math. Surveys **58** (2003), 173–174.

[2] M. Akcoglu, L. Sucheston, *On operator convergence in Hilbert space and in Lebesgue space*, Period. Math. Hungar. **72**, 235–244.

[3] N. I. Akhieser, I. M. Glazman, *Theory of Linear Operators in Hilbert Space*, Dover Publications, Inc., New York, 1993.

[4] J. Alber, *Von der Struktur implementierter Halbgruppen mit Anwendungen in der Stabilitätstheorie*, diploma thesis, Univerity of Tübingen, 1999.

[5] J. Alber, *On implemented semigroups*, Semigroup Forum **63** (2001), 371–386.

[6] G. R. Allan, T. J. Ransford, *Power-dominated elements in a Banach algebra*, Studia Math. **94** (1989), 63–79.

[7] G. R. Allan, A. G. Farrell, T. J. Ransford, *A Tauberian theorem arising in operator theory*, Bull. London Math. Soc. **19** (1987), 537–545.

[8] W. Arendt, *Semigroups and evolution equations: Functional calculus, regu-larity and kernel estimates*. In: Evolutionary Equations **Vol. I**. Handbook of Differential Equations. C. M. Dafermos, E. Feireisl eds., Elsevier, Amster-dam 2004, pp. 1–85.

[9] W. Arendt, C. J. K. Batty, *Tauberian theorems and stability of one-parameter semigroups*, Trans. Amer. Math. Soc. **306** (1988), 837–852.

[10] W. Arendt, C. J. K. Batty, M. Hieber, and F. Neubrander, *Vector-valued Laplace Transforms and Cauchy Problems*, Monographs in Mathematics, vol. 96, Birkhäuser, Basel, 2001.

[11] A. Avila, G. Forni, *Weak mixing for interval exchange transformations and translation flows*, Ann. of Math. (2) **165** (2007), 637–664.

[12] C. Badea, V. Müller, *On weak orbits of operators*, Topology Appl. **156** (2009), 1381–1385.

[13] C. Badea, S. Grivaux, *Size of the peripheral point spectrum under power or resolvent growth conditions*, J. Funct. Anal. **246** (2007), 302–329.

[14] C. Badea, S. Grivaux, *Unimodular eigenvalues, uniformly distributed sequences and linear dynamics*, Adv. Math. **211** (2007), 766–793.

[15] H. Bart, S. Goldberg, *Characterizations of almost periodic strongly continuous groups and semigroups*, Math. Ann. **236** (1978), 105–116.

[16] W. Bartoszek, *On the residuality of mixing by convolutions probabilities*, Israel J. Math. **80** (1992), 183–193.

[17] W. Bartoszek, B. Kuna, *Strong mixing Markov semigroups on C_1 are meager*, Colloq. Math. **105** (2006), 311–317.

[18] W. Bartoszek, B. Kuna, *On residualities in the set of Markov operators on C_1*, Proc. Amer. Math. Soc. **133** (2005), 2119–2129.

[19] A. Batkai, T. Eisner, and Yu. Latushkin, *The spectral mapping property of delay semigroups*, Compl. Anal. Oper. Theory **2** (2008), 273–283.

[20] A. Batkai, K. Engel, J. Prüss, R. Schnaubelt, *Polynomial stability of operator semigroups*, Math. Nachr. **279** (2006), 1425–1440.

[21] C. J. K. Batty, *Asymptotic behaviour of semigroups of linear operators*, in: J. Zemánek (ed.), Functional Analysis and Operator Theory, Banach Center Publications **30** (1994), 35–52.

[22] C. J. K. Batty, *Spectral conditions for stability of one-parameter semigroups*, J. Diff. Equ. **127** (1996), 87–96.

[23] C. J. K. Batty, R. Chill, and J. M. A. M. van Neerven, *Asymptotic behaviour of C_0-semigroups with bounded local resolvents*, Math. Nachr. **219** (2000), 65–83.

[24] C. J. K. Batty, R. Chill, and Yu. Tomilov, *Strong stability of bounded evolution families and semigroups*, J. Funct. Anal. **193** (2002), 116–139.

[25] C. J. K. Batty, T. Duyckaerts, *Non-uniform stability for bounded semi-groups on Banach spaces*, J. Evol. Equ. **8** (2008), 765–780.

[26] C. J. K. Batty, J. van Neerven, F. Räbiger, *Local spectra and individual stability of uniformly bounded C_0-semigroups*, Trans. Amer. Math. Soc. **350** (1998), 2071–2085.

[27] C. J. K. Batty, V. Q. Phóng, *Stability of individual elements under one-parameter semigroups*, Trans. Amer. Math. Soc. **322** (1990), 805–818.

[28] C. J. K. Batty, D. W. Robinson, *Positive one-parameter semigroups on ordered Banach spaces*, Acta Appl. Math. **2** (1984), 221–296.

[29] N.-E. Benamara, N. Nikolski, *Resolvent tests for similarity to a normal operator*, Proc. London Math. Soc. **78** (1999), 585–626.

[30] H. Bercovici, *On the iterates of completely nonunitary contractions*, Topics in Operator Theory: Ernst D. Hellinger memorial volume, Operator Theory Adv. Appl., Birkhäuser, Basel, Vol. 48 (1990), 185–188.

[31] D. Berend, M. Lin, J. Rosenblatt, and A. Tempelman, *Modulated and subsequential ergodic theorems in Hilbert and Banach spaces*, Erg. Theory Dynam. Systems **22** (2002), 1653–1665.

[32] S. Bergman, *Sur les fonctions orthogonales de plusieurs variables complexes avec les applications à la théorie des fonctions analytiques*, Gauthier–Villars, Paris, 1947.

[33] J. F. Berglund, H. D. Junghenn, and P. Milnes, *Analysis on Semigroups*, Canadian Mathematical Society Series of Monographs and Advanced Texts, John Wiley & Sons Inc., New York, 1989.

[34] C. Bluhm, *Liouville numbers, Rajchman measures, and small Cantor sets*, Proc. Amer. Math. Soc. **128** (2000), 2637–2640.

[35] A. Bobrowski, *A note on convergence of semigroups*, Ann. Polon. Math. **69** (1998), 107–127.

[36] A. Bobrowski, *Asymptotic behaviour of a Feller evolution family involved in the Fisher–Wright model*, Adv. App. Prob. **40** (2008), 734–758.

[37] A. Bobrowski, M. Kimmel, *Asymptotic behaviour of an operator exponential related to branching random walk models of DNA repeats*, J. Biological Systems **7** (1999), 33–43.

[38] A. Bobrowski, M. Kimmel, *Dynamics of the life history of a DNA-repeat sequence*, Archives Control Sciences **9** (1999), 57–67.

[39] A. Borichev, Yu. Tomilov, *Optimal polynomial decay of functions and operator semigroups*, Math. Ann. **347** (2010), 455–478.

[40] N. Borovykh, D. Drissi, M. N. Spijker, *A note about Ritt's condition, related resolvent conditions and power bounded operators*, Numer. Funct. Anal. Optim. **21** (2000), 425–438.

[41] P. Brenner, V. Thomée, *On rational approximations of semigroups*, SIAM J. Numer. Anal. **16** (1979), 683–694.

[42] P. Brenner, V. Thomée, and L. B. Wahlbin, *Besov Spaces and Applications to Difference Methods for Initial Value Problems*, Springer Verlag, Berlin, 1975.

[43] L. de Branges, J. Rovnyak, *The existence of invariant subspaces*, Bull. Amer. Math. Soc. **70** (1964), 718–721, and **71** (1965), 396.

[44] P. L. Butzer, U. Westphal, *On the Cayley transform and semigroup operators*, Hilbert space operators and operator algebras. Proceedings of an International Conference held at Tihany, 14–18 September 1970. Edited by Béla

Sz.-Nagy. Colloquia Mathematica Societatis János Bolyai, Vol. 5. North-Holland Publishing Co., Amsterdam-London, 1972.

[45] J. A. van Casteren, *Operators similar to unitary or selfadjoint ones*, Pacific J. Math. **104** (1983), 241–255.

[46] J. A. van Casteren, *Boundedness properties of resolvents and semigroups of operators*, in Linear Operators (Warsaw, 1994), 59–74, Banach Center Publ., 38, Polish Acad. Sci., Warsaw, 1997.

[47] C. Chicone, Yu. Latushkin, *Evolution semigroups in dynamical systems and differential equations*, Mathematical Surveys and Monographs, vol. 70, American Mathematical Society, Providence, 1999.

[48] R. Chill, *Tauberian theorems for vector-valued Fourier and Laplace transforms*, Stud. Math. **128** (1998), 55–69.

[49] R. Chill, Yu. Tomilov, *Stability of C_0-semigroups and geometry of Banach spaces*, Math. Proc. Cambridge Phil. Soc. **135** (2003), 493–511.

[50] R. Chill, Yu. Tomilov, *Stability of operator semigroups: ideas and results*, Perspectives in operator theory, 71–109, Banach Center Publ., 75, Polish Acad. Sci., Warsaw, 2007.

[51] J. R. Choksi, M. G. Nadkarni, *Baire category in spaces of measures, unitary operators, and transformations*, Proc. Int. Conference on Invariant Subspaces and Allied Topics (1986), Narosa Publishers, New Delhi.

[52] J. B. Conway, *A Course in Functional Analysis*. Second edition, Graduate Texts in Mathematics 96, Springer-Verlag, New York, 1990.

[53] I. P. Cornfeld, S. V. Fomin, and Ya. G. Sinai, *Ergodic Theory*, Grundlehren der mathematischen Wissenschaften 245, Springer-Verlag, 1982.

[54] H. G. Dales, P. Aiena, J. Eschmeier, K. Laursen, G. A. Willis, *Introduction to Banach Algebras, Operators, and Harmonic Analysis*, London Mathematical Society Student Texts, Cambridge University Press, Cambridge, 2003.

[55] R. Datko, *Extending a theorem of A. M. Liapunov to Hilbert space*, J. Math. Anal. Appl. **32** (1970), 610–616.

[56] K. R. Davidson, *Polynomially bounded operators, a survey*, Operator algebras and applications (Samos, 1996), 145–162, NATO Adv. Sci. Inst. Ser. C Math. Phys. Sci., 495, Kluwer Acad. Publ., Dordrecht, 1997.

[57] E. B. Davies, *One-Parameter Semigroups*, Academic Press, London–New York–San Fransisco, 1980.

[58] D. Deckard, C. Pearcy, *Another class of invertible operators without square roots*, Proc. Amer. Math. Soc. **14** (1963), 445–449.

[59] T. de la Rue, J. de Sam Lazaro, *The generic transformation can be embedded in a flow* (French), Ann. Inst. H. Poincaré, Prob. Statist. **39** (2003), 121–134.

[60] R. de Laubenfels, *Inverses of generators*, Proc. Amer. Math. Soc. **104** (1988), 443–448.

[61] R. Derndinger, R. Nagel, G. Palm, *13 Lectures on Ergodic Theory. Functional Analytic View.* Manuscript, 1987.

[62] J. Dugundji, *Topology*, Allyn and Bacon, 1966.

[63] N. Dunford, J. T. Schwartz, *Linear Operators. I.*, Interscience Publishers, Inc., New York; Interscience Publishers, Ltd., London 1958.

[64] T. Eisner, *Polynomially bounded C_0-semigroups*, Semigroup Forum **70** (2005), 118–126.

[65] T. Eisner, *A "typical" contraction is unitary*, L'Enseignement Mathématique, to appear.

[66] T. Eisner, *Embedding operators into strongly continuous semigroups*, Arch. Math. (Basel) **92** (2009), 451–460.

[67] T. Eisner, B. Farkas, R. Nagel, and A. Serény, *Weakly and almost weakly stable C_0-semigroups*, Inter. J. Dyn. Syst. Diff. Eq. **1** (2007), 44–57.

[68] T. Eisner, B. Farkas, M. Haase, and R. Nagel, *Ergodic Theory: an Operator Theoretic Approach*, book manuscript, in preparation.

[69] T. Eisner, A. Serény, *Category theorems for stable operators on Hilbert spaces*, Acta Sci. Math. (Szeged) **74** (2008), 259–270.

[70] T. Eisner, A. Serény, *Category theorems for stable semigroups*, Erg. Theory Dynam. Systems **29** (2009), 487–494.

[71] T. Eisner, A. Serény, *On the weak analogue of the Trotter–Kato theorem*, Taiwanese J. Math., to appear.

[72] T. Eisner, H. Zwart, *Continuous-time Kreiss resolvent condition on infinite-dimensional spaces*, Math. Comp. **75** (2006), 1971–1985.

[73] T. Eisner, H. Zwart, *A note on polynomially bounded C_0-semigroups*, Semigroup Forum **75** (2007), 438–445.

[74] T. Eisner, H. Zwart, *The growth of a C_0-semigroup characterised by its cogenerator*, J. Evol. Equ. **8** (2008), 749–764.

[75] E. Yu. Emel'yanov, *Non-spectral Asymptotic Analysis of One-parameter Operator Semigroups*, Operator Theory: Advances and Applications, vol. 173, Birkhäuser Verlag, Basel, 2007.

[76] R. Emilion, *Mean-bounded operators and mean ergodic theorems*, J. Funct. Anal. **61** (1985), 1–14.

[77] Z. Emirsajlow, S. Townley, *On application of the implemented semigroup to a problem arising in optimal control*, Internat. J. Control **78** (2005), 298–310.

[78] K.-J. Engel, R. Nagel, *One-parameter Semigroups for Linear Evolution Equations*, Graduate Texts in Mathematics, vol. 194, Springer-Verlag, New York, 2000.

[79] K.-J. Engel, R. Nagel, *A Short Course on Operator Semigroups*, Universitext, Springer-Verlag, New York, 2006.

[80] J. Esterle, E. Strouse, and F. Zouakia, *Theorems of Katznelson-Tzafriri type for contractions*, J. Funct. Anal. **94** (1990), 273–287.

[81] J. Esterle, E. Strouse, and F. Zouakia, *Stabilité asymptotique de certains semi-groupes d'opérateurs et idéaux primaires de $L^1(\mathbb{R}_+)$*, J. Operator Theory **28** (1992), 203–227.

[82] A. Iwanik, *Baire category of mixing for stochastic operators*, Measure theory (Oberwolfach, 1990), Rend. Circ. Mat. Palermo (2) Suppl. **28** (1992), 201–217.

[83] S. Ferenczi, C. Holton, L. Q. Zamboni, *Joinings of three-interval exchange transformations*, Ergodic Theory Dynam. Systems **25** (2005), 483–502.

[84] M. J. Fisher, *The embeddability of an invertible measure*, Semigroup Forum **5** (1972/73), 340–353.

[85] C. Foiaş, *A remark on the universal model for contractions of G. C. Rota*, Com. Acad. R. P. Romîne **13** (1963), 349–352.

[86] S. R. Foguel, *Powers of a contraction in Hilbert space*, Pacific J. Math. **13** (1963), 551–562.

[87] S. R. Foguel, *A counterexample to a problem of Sz. Nagy*, Proc. Amer. Math. Soc. **15** (1964), 788–790.

[88] S. R. Foguel, *The Ergodic Theory of Markov Processes*, Van Nostrand Mathematical Studies, No. 21. Van Nostrand Reinhold Co., New York-Toronto, Ont.-London, 1969.

[89] V. P. Fonf, M. Lin, P. Wojtaszczyk, *Ergodic characterization of reflexivity of Banach spaces*, J. Funct. Anal. **187** (2001), 146–162.

[90] K. Frączek, M. Lemańczyk, *On mild mixing of special flows over irrational rotations under piecewise smooth functions*, Ergodic Theory Dynam. Systems **26** (2006), 719–738.

[91] K. Frączek, M. Lemańczyk, E. Lesigne, *Mild mixing property for special flows under piecewise constant functions*, Discrete Contin. Dyn. Syst. **19** (2007), 691–710.

[92] H. Furstenberg, *Recurrence in Ergodic Theory and Combinatorial Number Theory*, Princeton University Press, Princeton, New Jersey, 1981.

[93] H. Furstenberg, B. Weiss, *The finite multipliers of infinite ergodic transformations*, The structure of attractors in dynamical systems (Proc. Conf., North Dakota State Univ., Fargo, N.D., 1977), pp. 127–132, Lecture Notes in Math., 668, Springer, Berlin, 1978.

[94] L. Gearhart, *Spectral theory for contraction semigroups on Hilbert spaces*, Trans. Amer. Math. Soc. **236** (1978), 385–394.

[95] I. M. Gelfand, *Zur Theorie der Charaktere der abelschen topologischen Gruppen*, Rec. Math. N. S. (Mat. Sb.) **9 (51)** (1941), 49–50.

[96] E. Glasner, *Ergodic Theory via Joinings*, Mathematical Surveys and Monographs, Amer. Math. Soc., Providence, 2003.

[97] K. de Leeuw, I. Glicksberg, *Applications of almost periodic compactifications*, Acta Math. **105** (1961), 63–97.

[98] J. A. Goldstein, *An asymptotic property of solutions of wave equations*, Proc. Amer. Math. Soc. **23** (1969) 359–363.

[99] J. A. Goldstein, *An asymptotic property of solutions of wave equations. II*, J. Math. Anal. Appl. **32** (1970), 392–399.

[100] J. A. Goldstein, *Semigroups of Linear Operators and Applications*, Oxford Mathematical Monographs, Oxford University Press, New York, 1985.

[101] J. A. Goldstein, *Asymptotics for bounded semigroups on Hilbert space*, Aspects of positivity in functional analysis (Tübingen, 1985), 49–62, North-Holland Math. Stud., 122, North-Holland, Amsterdam, 1986.

[102] J. A. Goldstein, *Applications of operator semigroups to Fourier analysis*, Semigroup Forum **52** (1996), 37–47.

[103] J. A. Goldstein, M. Wacker, *The energy space and norm growth for abstract wave equations*, Appl. Math. Lett. **16** (2003), 767–772.

[104] A. M. Gomilko, *Conditions on the generator of a uniformly bounded C_0-semigroup*, Functional Analysis and Appl. **33** (1999), 294–296.

[105] A. M. Gomilko, *The Cayley transform of the generator of a uniformly bounded C_0-semigroup of operators*, Ukrainian Math. J. **56** (2004), 1212–1226.

[106] A. M. Gomilko, J. Zemánek,
a) *On the uniform Kreiss resolvent condition*, Funct. Anal. Appl. **42** (2008), 230–233.
b) *On the strong Kreiss resolvent condition in the Hilbert space*, Operator Theory: Advances and Applications, vol. 190, 237–242, Birkhäuser Verlag, 2009.

[107] A. M. Gomilko, H. Zwart, *The Cayley transform of the generator of a bounded C_0-semigroup*, Semigroup Forum **74** (2007), 140–148.

[108] A. M. Gomilko, H. Zwart, Yu. Tomilov, *On the inverse of the generator of a C_0-semigroup*, Mat. Sb., **198** (2007), 35–50.

[109] G. R. Goodson, J. Kwiatkowski, M. Lemańczyk, P. Liardet, *On the multiplicity function of ergodic group extensions of rotations*, Studia Math. **102** (1992), 157–174.

[110] G. Greiner, R. Nagel, *On the stability of strongly continuous semigroups of positive operators on $L^2(\mu)$*, Ann. Scuola Norm. Sup. Pisa Cl. Sci. (4) **10** (1983), 257–262.

[111] G. Greiner, M. Schwarz, *Weak spectral mapping theorems for functional-differential equations* J. Differential Equations **94** (1991), 205–216.

[112] U. Groh, F. Neubrander, *Stabilität starkstetiger, positiver Operatorhalbgruppen auf C^*-Algebren*, Math. Ann. **256** (1981), 509–516.

[113] B. Z. Guo, H. Zwart, *On the relation between stability of continuous- and discrete-time evolution equations via the Cayley transform*, Integral Equations Oper. Theory **54** (2006), 349–383.

[114] M. Haase, *The Functional Calculus for Sectorial Operators*, Operator Theory: Advances and Applications, vol. 169, Birkhäuser Verlag, Basel, 2006.

[115] M. Haase, *Functional calculus for groups and applications to evolution equations*, J. Evol. Equ. **7** (2007), 529–554.

[116] P. R. Halmos, *In general a measure preserving transformation is mixing*, Ann. Math. **45** (1944), 786–792.

[117] P. R. Halmos, *Introduction to Hilbert Space and the Theory of Spectral Multiplicity*, Chelsea Publishing Company, New York, 1951.

[118] P. R. Halmos, *Lectures on Ergodic Theory*, Chelsea Publishing Co., New York 1960.

[119] P. R. Halmos, *What does the spectral theorem say?*, Amer. Math. Monthly **70** (1963), 241–247.

[120] P. Halmos, *On Foguel's answer to Nagy's question*, Proc. Amer. Math. Soc. **15** (1964), 791–793.

[121] P. R. Halmos, *A Hilbert Space Problem Book*, D. Van Nostrand Co., Inc., Princeton, N.J.-Toronto, Ont.-London 1967.

[122] P. R. Halmos, G. Lumer, J. J. Schäffer, *Square roots of operators*, Proc. Amer. Math. Soc. **4** (1953), 142–149.

[123] G. H. Hardy, *Divergent Series*, Oxford, Clarendon Press, 1949.

[124] H. Hedenmalm, B. Korenblum, K. Zhu, *Theory of Bergman spaces*, Graduate Texts in Mathematics, Springer-Verlag, New York, 2000.

[125] I. Herbst, *The spectrum of Hilbert space semigroups*, J. Operator Theory **10** (1983), 87–94.

[126] R. Hersh, T. Kato, *High-accuracy stable difference schemes for well-posed initial value problems*, SIAM J. Numer. Anal., **16** (1979), 670–682.

[127] E. Hewitt, K. Ross, *Abstract harmonic analysis. Vol. I.*, Springer-Verlag, Berlin–New York, 1979.

[128] H. Heyer, *Probability Measures on Locally Compact Groups*, Ergebnisse der Mathematik und ihrer Grenzgebiete, Vol. 94, Springer–Verlag, Berlin-New York, 1977.

[129] F. Hiai, *Weakly mixing properties of semigroups of linear operators*, Kodai Math. J. **1** (1978), 376–393.

[130] F. L. Huang, *Characteristic conditions for exponential stability of linear dynamical systems in Hilbert spaces*, Ann. Differential Equations **1** (1985), 43–56.

[131] S.-Z. Huang, *An equivalent description of non-quasianalyticity through spectral theory of C_0-groups*, J. Operator Theory **32** (1994), 299–309.

[132] S.-Z. Huang, *A local version of Gearhart's theorem*, Semigroup Forum **58** (1999), 323–335.

[133] A. Izzo, *A functional analysis proof of the existence of Haar measure on locally compact abelian groups*, Proc. Amer. Math. Soc. **115** (1992), 581–583.

[134] K. Jacobs, *Periodizitätseigenschaften beschränkter Gruppen im Hilbertschen Raum*, Math. Z. **61** (1955), 340–349.

[135] B. Jamison, *Eigenvalues of modulus 1*, Proc. Amer. Math. Soc. **16** (1965), 375–377.

[136] J. K. Jones, V. Kuftinec, *A note on the Blum–Hanson theorem*, Proc. Amer. Math. Soc. **30** (1971), 202–203.

[137] L. K. Jones, M. Lin, *Ergodic theorems of weak mixing type*, Proc. Amer. Math. Soc. **57** (1976), 50–52.

[138] L. K. Jones, M. Lin, *Unimodular eigenvalues and weak mixing*, J. Funct. Anal. **35** (1980), 42–48.

[139] M. A. Kaashoek, S. M. Verduyn Lunel, *An integrability condition on the resolvent for hyperbolicity of the semigroup*, J. Diff. Eq. **112** (1994), 374–406.

[140] C. Kaiser, L. Weis, *A perturbation theorem for operator semigroups in Hilbert spaces*, Semigroup Forum **67** (2003), 63–75.

[141] A. Katok, *Interval exchange transformations and some special flows are not mixing*, Israel J. Math. **35** (1980), 301–310.

[142] A. Katok, *Approximation and Generity in Abstract Ergodic Theory*, Notes 1985.

[143] A. Katok, B. Hasselblatt, *Introduction to the Modern Theory of Dynamical Systems*, Encyclopedia of Mathematics and its Applications, vol. 54, Cambridge University Press, Cambridge, 1995.

[144] A. B. Katok, A. M. Stepin, *Approximations in ergodic theory*, Uspehi Mat. Nauk **22** (1967), 81–106.

[145] W.-P. Katz, *Funktionalkalkül und Asymptotik von Kontraktionen auf Hilberträumen*, diploma thesis, University of Tübingen, 1994.

[146] Y. Katznelson, *An introduction to harmonic analysis*, Second corrected edition. Dover Publications, Inc., New York, 1976.

[147] Y. Katznelson, L. Tzafriri, *On power bounded operators*, J. Funct. Anal. **68** (1986), 313–328.

[148] L. Kérchy, J. van Neerven, *Polynomially bounded operators whose spectrum on the unit circle has measure zero*, Acta Sci. Math. (Szeged) **63** (1997), 551–562.

[149] J. L. King, *The generic transformation has roots of all orders*, Colloq. Math. **84/85** (2000), 521–547.

[150] A. V. Kiselev, *On the resolvent estimates for the generators of strongly continuous groups in the Hilbert spaces*, Operator methods in ordinary and partial differential equations (Stockholm, 2000), 253–266, Oper. Theory Adv. Appl. **132**, Birkhäuser, Basel, 2002.

[151] T. W. Körner, *On the theorem of Ivashev-Musatov III*, Proc. London Math. Soc. **53** (1986), 143–192.

[152] H. Komatsu, *Fractional powers of operators*, Pacific J. Math. **19** (1966), 285–346.

[153] M. Kramar, E. Sikolya, *Spectral properties and asymptotic periodicity of flows in networks*, Math. Z. **249** (2005), 139–162.

[154] U. Krengel, *Ergodic Theorems*, de Gruyter Studies in Mathematics, de Gruyter, Berlin, 1985.

[155] S. Król, *A note on approximation of semigroups of contractions on Hilbert spaces*, Semigroup Forum **79** (2009), 369–376.

[156] R. Kühne, *On weak mixing for semigroups*, Ergodic theory and related topics (Vitte, 1981), 133–139, Math. Res. **12**, Akademie-Verlag, Berlin, 1982.

[157] R. Kühne, *Weak mixing for representations of semigroups on a Banach space*, Proceedings of the conference on ergodic theory and related topics, II (Georgenthal, 1986), 108–112, Teubner-Texte Math. **94**, Teubner, Leipzig, 1987.

[158] F. Kühnemund, *A Hille-Yosida theorem for bi-continuous semigroups*, Semigroup Forum **67** (2003), 205–225.

[159] J. Kulaga, *On self-similarity problem for smooth flows on orientable surfaces*, submitted, 2009.

[160] A. Lasota, J. Myjak, *Generic properties of stochastic semigroups*, Bull. Polish Acad. Sci. Math. **40** (1992), 283–292.

[161] Yu. Latushkin, F. Räbiger, *Operator valued Fourier multipliers and stability of strongly continuous semigroups*, Integral Equations Operator Theory **51** (2005), 375–394.

[162] Yu. Latushkin, R. Shvydkoy, *Hyperbolicity of semigroups and Fourier multipliers*, Systems, approximation, singular integral operators, and related topics (Bordeaux, 2000), 341–363, Oper. Theory Adv. Appl. **129**, Birkhäuser, Basel, 2001.

[163] P. D. Lax, *Functional Analysis*, Pure and Applied Mathematics, Wiley-Interscience, New York, 2002.

[164] P. D. Lax, R. S. Phillips, *Scattering Theory*, Pure and Applied Mathematics, Academic Press, New York–London 1967.

[165] M. Lemańczyk, C. Mauduit, *Ergodicity of a class of cocycles over irrational rotations*, J. London Math. Soc. **49** (1994), 124–132.

[166] A. M. Liapunov, *Stability of motion*, Academic Press, New York–London, 1966. Ph.D. thesis, Kharkov, 1892.

[167] M. Lin, *Mixing for Markov operators*, Z. Wahrscheinlichkeitstheorie und Verw. Gebiete **19** (1971), 231–242.

[168] M. Lin, *On the uniform ergodic theorem*, Proc. Amer. Math. Soc. **43** (1974), 337–340.

[169] M. Lin, *On the uniform ergodic theorem. II*, Proc. Amer. Math. Soc. **46** (1974), 217–225.

[170] M. Lin, *Quasi-compactness and uniform ergodicity of positive operators*, Israel J. Math. **29** (1978), 309–311.

[171] D. A. Lind, *A counterexample to a conjecture of Hopf*, Duke Math. J. **42** (1975), 755–757.

[172] Ch. Lubich, O. Nevanlinna, *On resolvent conditions and stability estimates*, BIT **31** (1991), 293–313.

[173] Z.-H. Luo, B.-Z. Guo, and O. Morgul, *Stability and stabilization of infinite dimensional systems with applications*, Springer-Verlag, London, 1999.

[174] R. Lyons, *Characterization of measures whose Fourier-Stiltjes transforms vanish at infinity*, Bull. Amer. Math. Soc. **10** (1984), 93–96.

[175] R. Lyons, *Fourier-Stieltjes coefficients and asymptotic distribution modulo 1*, Annals Math. **122** (1985), 155–170.

[176] R. Lyons, *The measure of non-normal sets*, Invent. Math. **83** (1986), 605–616.

[177] R. Lyons, *The size of some classes of thin sets*, Studia Math. **86** (1987), 59–78.

[178] R. Lyons, *Topologies on measure spaces and the Radon-Nikodym property*, Studia Math. **91** (1988), 125–129.

[179] R. Lyons, *Seventy years of Rajchman measures*, J. Fourier Anal. Appl. (1995), 363–377.

[180] Yu. I. Lyubich, Vũ Quôc Phóng, *Asymptotic stability of linear differential equations in Banach spaces*, Studia Math. **88** (1988), 37–42.

[181] Yu. I. Lyubich, Vũ Quôc Phóng, *A spectral criterion for asymptotic almost periodicity for uniformly continuous representations of abelian semigroups*, J. Soviet Math. **49** (1990), 1263–1266.

[182] W. Maak, *Periodizitätseigenschaften unitärer Gruppen in Hilberträumen*, Math. Scand. **2** (1954), 334–344.

[183] M. Malejki, C_0-*groups with polynomial growth*, Semigroup Forum **63**, (2001), 305–320.

[184] D.E. Menshov, *Sur l'unicité du développement trigonométrique*, C. R. Acad. Sc. Paris, Sér. A-B **163** (1916), 433–436.

[185] H. Milicer-Grużewska, *Sur les fonctions à variation bornée et à l'écart Hadamardien nul*, C. R. Soc. Sci. Varsovie **21** (1928), 67–78.

[186] L. A. Monauni, *On the abstract Cauchy problem and the generation problem for semigroups of bounded operators*, Control Theory Centre Report No. 90, Warwick, 1980.

[187] A. Montes-Rodrígez, J. Sánchez-Álvarez, and J. Zemánek, *Uniform Abel-Kreiss boundedness and the extremal behaviour of the Volterra operator*, Proc. London Math. Soc. **91** (2005), 761–788.

[188] V. Müller, *Local spectral radius formula for operators in Banach spaces*, Czechoslovak Math. J. **38 (133)** (1988), 726–729.

[189] V. Müller, *Orbits, weak orbits and local capacity of operators*, Integral Equations Operator Theory **41** (2001), 230–253.

[190] V. Müller, *Power bounded operators and supercyclic vectors*, Proc. Amer. Math. Soc. **131** (2003), 3807–3812.

[191] V. Müller, *Spectral theory of linear operators and spectral systems in Banach algebras*, second edition. Operator Theory: Advances and Applications, vol. 139, Birkhäuser Verlag, Basel, 2007.

[192] V. Müller, Yu. Tomilov, *Quasisimilarity of power bounded operators and Blum–Hanson property*, J. Funct. Anal. **246** (2007), 285–300.

[193] M. G. Nadkarni, *Spectral Theory of Dynamical Systems*, Birkhäuser Advanced Texts: Basel Textbooks, Birkhäuser Verlag, Basel, 1998.

[194] R. Nagel, *Mittelergodische Halbgruppen linearer Operatoren*, Ann. Inst. Fourier (Grenoble) **23** (1973), 75–87.

[195] R. Nagel, *Ergodic and mixing properties of linear operators*, Proc. Roy. Irish Acad. Sect. A. **74** (1974), 245–261.

[196] R. Nagel (ed.), *One-parameter Semigroups of Positive Operators*, Lecture Notes in Mathematics, vol. 1184, Springer-Verlag, Berlin, 1986.

[197] R. Nagel, S.-Z. Huang, *Spectral mapping theorems for C_0-groups satisfying non-quasianalytic growth conditions*, Math. Nachr. **169** (1994), 207–218.

[198] R. Nagel, F. Räbiger, *Superstable operators on Banach spaces*, Israel J. Math. **81** (1993), 213–226.

[199] R. Nagel, A. Rhandi, *Positivity and Liapunov's stability conditions for linear systems*, Adv. Math. Sci. Appl. **3** (1993/94), Special Issue, 33–41.

[200] B. Nagy, J. Zemánek, *A resolvent condition implying power boundedness*, Studia Math. **134** (1999), 143–151.

[201] I. Namioka, *Separate continuity and joint continuity*, Pacific J. Math. **57** (1974), 515–531.

[202] J. M. A. M. van Neerven, *On the orbits of an operator with spectral radius one*, Czechoslovak Math. J. **45 (120)** (1995), 495–502.

[203] J. M. A. M. van Neerven, *Exponential stability of operators and operator semigroups*, J. Funct. Anal. **130** (1995), 293–309.

[204] J. M. A. M. van Neerven, *The Asymptotic Behaviour of Semigroups of Linear Operators*, Operator Theory: Advances and Applications, vol. 88, Birkhäuser Verlag, Basel, 1996.

[205] J. von Neumann, *Eine Spektraltheorie für allgemeine Operatoren eines unitären Raumes*, Math. Nachrichten 4 (1951), 258–281.

[206] O. Nevanlinna, *Resolvent conditions and powers of operators*, Studia Math. **145** (2001), 113–134.

[207] E. W. Packel, *A semigroup analogue of Foguel's counterexample*, Proc. Amer. Math. Soc. **21** (1969), 240–244.

[208] T. W. Palmer, *Banach Algebras and the General Theory of *-Algebras I*, Algebras and Banach algebras, Encyclopedia of Mathematics and its Applications, vol. 49, Cambridge University Press, Cambridge, 1994.

[209] A. Pazy, *Semigroups of Linear Operators and Applications to Partial Differential Equations*, Springer–Verlag, 1983.

[210] G. K. Pedersen, *Analysis now*, Graduate Texts in Mathematics, 118. Springer-Verlag, New York, 1989.

[211] V. V. Peller, *Estimates of operator polynomials in the space L^p with respect to the multiplicative norm*, J. Math. Sciences **16** (1981), 1139–1149.

[212] K. Petersen, *Ergodic Theory*, Cambridge Studies in Advanced Mathematics, Cambridge University Press, 1983.

[213] G. Pisier, *Similarity problems and completely bounded maps*, Lecture Notes in Mathematics, 1618. Springer-Verlag, Berlin, 1996.

[214] G. Pisier, *A polynomially bounded operator on Hilbert space which is not similar to a contraction*, J. Amer. Math. Soc. **10** (1997), 351–369.

[215] J. Prüss, *On the spectrum of C_0-semigroups*, Trans. Amer. Math. Soc. **284** (1984), 847–857.

[216] A. Rajchman, *Sur une classe de fonctions à variation bornée*, C. R. Acad. Sci. Paris **187** (1928), 1026–1028.

[217] A. Rajchman, *Une classe de séries trigonométriques qui convergent presque partout vers zéro*, Math. Ann. **101** (1929), 686–700.

[218] T. Ransford, *Eigenvalues and power growth*, Israel J. Math. **146** (2005), 93–110.

[219] T. Ransford, M. Roginskaya, *Point spectra of partially power-bounded operators*, J. Funct. Anal. **230** (2006), 432–445.

[220] F. Riesz, B. Sz.-Nagy, *Functional Analysis*, Frederick Ungar Publishing Co., 1955.

[221] V. A. Rohlin, *A "general" measure-preserving transformation is not mixing*, Doklady Akad. Nauk SSSR **60** (1948), 349–351.

[222] S. Rolewicz, *On uniform N–equistability*, J. Math. Anal. Appl. **115** (1986), 434–441.

[223] M. Rosenblum, J. Rovnyak, *Topics in Hardy Classes and Univalent Functions*, Birkhäuser, Basel, 1994.

[224] W. M. Ruess, W. H. Summers, *Weak asymptotic almost periodicity for semi-groups of operators*, J. Math. Anal. Appl. **164** (1992), 242–262.

[225] W. Rudin, *Functional Analysis*, 2nd edition. International Series in Pure and Applied Mathematics, McGraw-Hill, New York, 1991.

[226] H. H. Schaefer, *Topological Vector Spaces*. Third printing corrected. Graduate Texts in Mathematics, Springer-Verlag, 1971.

[227] H. H. Schaefer, *Banach Lattices and Positive Operators*, Springer-Verlag, 1974.

[228] D. Scheglov, *Absence of mixing for smooth flows on genus two surfaces*, J. Mod. Dyn. **3** (2009), 13–34.

[229] B. Simon, *Operators with singular continuous spectrum. I. General operators*, Ann. of Math. **141** (1995), 131–145.

[230] S.-Y. Shaw, *Growth order and stability of semigroups and cosine functions*, J. Math. Anal. Appl. **357** (2009), 340–348.

[231] D.-H. Shi, D.-X. Feng, *Characteristic conditions on the generator of C_0-semigroups in a Hilbert space*, J. Math. Anal. Appl. **247** (2000), 356–376.

[232] A. L. Shields, *On Möbius bounded operators*, Acta Sci. Math. (Szeged) **40** (1978), 371–374.

[233] M. N. Spijker, S. Tracogna, B. D. Welfert, *About the sharpness of the stability estimates in the Kreiss matrix theorem*, Math. Comp. **72** (2003), 697–713.

[234] G. M. Sklyar, V. Ya. Shirman, *Asymptotic stability of a linear differential equation in a Banach space* (Russian), Teor. Funktsiy Funktsional. Anal. i Prilozhen. **37** (1982), 127–132.

[235] A. M. Stepin, A. M. Eremenko, *Nonuniqueness of an inclusion in a flow and the vastness of a centralizer for a generic measure-preserving transformation*, Mat. Sb. **195** (2004), 95–108; translation in Sb. Math. **195** (2004), 1795–1808.

[236] B. Sz.-Nagy, *On uniformly bounded linear transformations in Hilbert space*, Acta Univ. Szeged. Sect. Sci. Math. **11** (1947), 152–157.

[237] B. Sz.-Nagy, C. Foiaş, *Sur les contractions de l'espace de Hilbert. IV*, Acta Sci. Math. Szeged **21** (1960), 251–259.

[238] B. Sz.-Nagy, C. Foiaş, *Harmonic Analysis of Operators on Hilbert Space*, North-Holland Publishing Co., 1970.

[239] M. Takesaki, *Theory of Operator Algebras I*, Springer-Verlag, 1979.

[240] T. Tao, *Poincaré's legacies, pages from year two of a mathematical blog. Part I*. American Mathematical Society, Providence, RI, 2009.

[241] D. Tsedenbayar, J. Zemánek, *Polynomials in the Volterra and Ritt operators*, Topological algebras, their applications, and related topics, 385–390, Banach Center Publ. **67**, Polish Acad. Sci., Warsaw, 2005.

[242] Yu. Tomilov, *On the spectral bound of the generator of a C_0-semigroup*, Studia Math. **125** (1997), 23–33.

[243] Yu. Tomilov, *A resolvent approach to stability of operator semigroups*, J. Operator Th. **46** (2001), 63–98.

[244] C. Ulcigrai, *Weak mixing for logarithmic flows over interval exchange transformations*, J. Mod. Dyn. **3** (2009), 35–49.

[245] V. M. Ungureanu, *Uniform exponential stability and uniform observability of time-varying linear stochastic systems in Hilbert spaces*, Recent advances in operator theory, operator algebras, and their applications, 287–306, Oper. Theory Adv. Appl., 153, Birkhäuser, Basel, 2005.

[246] P. Vitse, *Functional calculus under the Tadmor–Ritt condition, and free interpolation by polynomials of a given degree*, J. Funct. Anal. **210** (2004), 43–72.

[247] P. Vitse, *Functional calculus under Kreiss type conditions*, Math. Nachr. **278** (2005), 1811–1822.

[248] P. Vitse, *A band limited and Besov class functional calculus for Tadmor–Ritt operators*, Arch. Math. (Basel) **85** (2005), 374–385.

[249] Q. Ph. Vũ, *A short proof of the Y. Katznelson's and L. Tzafriri's theorem*, Proc. Amer. Math. Soc. **115** (1992), 1023–1024.

[250] Q. Ph. Vũ, *Theorems of Katznelson–Tzafriri type for semigroups of operators*, J. Funct. Anal. **119** (1992), 74–84.

[251] Q. Ph. Vũ, *Almost periodic and strongly stable semigroups of operators*, Linear operators (Warsaw, 1994), 401–426, Banach Center Publ., 38, Polish Acad. Sci., Warsaw, 1997.

[252] Q. Ph. Vũ, *On stability of C_0-semigroups*, Proc. Amer. Math. Soc. **129** (2001), 2871–2879.

[253] Q. Ph. Vũ, F. Yao, *On similarity to contraction semigroups in Hilbert space*, Semigroup Forum **56** (1998), 197–204.

[254] P. Walters, *An Introduction to Ergodic Theory*, Graduate Texts in Mathematics, Springer-Verlag, New York–Berlin, 1982.

[255] G. Weiss, *Weak L^p-stability of a linear semigroup on a Hilbert space implies exponential stability*, J. Diff. Eq. **1988**, 269–285.

[256] G. Weiss, *Weakly l^p-stable operators are power stable*, Int. J. Systems Sci. **20** (1989), 2323–2328.

[257] G. Weiss, *The resolvent growth assumption for semigroups on Hilbert spaces*, J. Math. Anal. Appl. **145** (1990), 154–171.

[258] L. Weis, V. Wrobel, *Asymptotic behaviour of C_0-semigroups on Banach spaces*, Proc. Amer. Math. Soc. **124** (1996), 3663–3671.

[259] G.-Q. Xu, D.-X. Feng, *On the spectrum determined growth assumption and the perturbation of C_0-semigroups*, Integral Equations Operator Theory **39** (2001), 363–376.

[260] K. Yosida, *Functional Analysis.* Fourth edition. Die Grundlehren der mathematischen Wissenschaften, Band 123. Springer-Verlag, New York-Heidelberg, 1974.

[261] A. C. Zaanen, *Continuity, integration and Fourier theory*, Universitext, Springer-Verlag, Berlin, 1989.

[262] J. Zabczyk, *A note on C_0-semigroups*, Bull. Acad. Polon. Sci. Ser. Math. Astr. Phys. **23** (1975), 895–898.

[263] H. Zwart, *Boundedness and strong stability of C_0-semigroups on a Banach space*, Ulmer Seminare 2003, pp. 380–383.

[264] H. Zwart, *On the estimate $\|(sI - A)^{-1}\| \leq M/\mathrm{Re}\, s$*, Ulmer Seminare 2003, pp. 384–388.

[265] H. Zwart, *Is A^{-1} an infinitesimal generator?*, Banach Center Publication **75** (2007), 303–313.

[266] H. Zwart, *Growth estimates for $exp(A^{-1}t)$ on a Hilbert space*, Semigroup Forum **74** (2007), 487–494.

List of Symbols

Index